Data Analytics and Machine Learning for Integrated Corridor Management

In an era defined by rapid urbanization and ever-increasing mobility demands, effective transportation management is paramount. This book takes readers on a journey through the intricate web of contemporary transportation systems, offering unparalleled insights into the strategies, technologies, and methodologies shaping the movement of people and goods in urban landscapes.

From the fundamental principles of traffic signal dynamics to the cutting-edge applications of machine learning, each chapter of this comprehensive guide unveils essential aspects of modern transportation management systems. Chapter by chapter, readers are immersed in the complexities of traffic signal coordination, corridor management, data-driven decision-making, and the integration of advanced technologies. Closing with chapters on modeling measures of effectiveness and computational signal timing optimization, the guide equips readers with the knowledge and tools needed to navigate the complexities of modern transportation management systems.

With insights into traffic data visualization and operational performance measures, this book empowers traffic engineers and administrators to design 21st-century signal policies that optimize mobility, enhance safety, and shape the future of urban transportation.

Data Analytics and Machine Learning for Integrated Corridor Management

Yashaswi Karnati, Dhruv Mahajan,
Tania Banerjee, Rahul Sengupta,
Clay Packard, Ryan Casburn,
Nithin Agarwal, Jeremy Dilmore,
Anand Rangarajan, and Sanjay Ranka

CRC Press is an imprint of the
Taylor & Francis Group, an **informa** business

First edition published 2025
by CRC Press
2385 NW Executive Center Drive, Suite 320, Boca Raton FL 33431

and by CRC Press
4 Park Square, Milton Park, Abingdon, Oxon, OX14 4RN

CRC Press is an imprint of Taylor & Francis Group, LLC

ISBN: 978-1-032-57464-6 (hbk)
ISBN: 978-1-032-57471-4 (pbk)
ISBN: 978-1-003-46030-5 (ebk)

DOI: 10.1201/9781003460305

Typeset in TexGyreTermesX font
by KnowledgeWorks Global Ltd.

Publisher's note: This book has been prepared from camera-ready copy provided by the authors.

This book is intended for pedagogical purposes and the authors have taken care to provide details based on their experience. However, the authors do not guarantee that the material in this book is accurate or will be effective in practice, nor are they responsible for any statement, material or formula that may have negative consequences, injuries, death etc. to the readers or the users of the book.

Dedication

Yashaswi Karnati: To all my friends, family, and mentors, whose unwavering support and guidance have made this journey possible.

Dhruv Mahajan: To my parents, mentors and friends who have always encouraged and supported me.

Tania Banerjee: I dedicate this book to my family.

Rahul Sengupta: To my family, mentors and friends, whose love, support and guidance made this possible.

Clay Packard: To my colleagues and friends at FDOT and throughout the transportation technology industry. Thank you for all your opportunities, collaboration, and support.

Ryan Casburn: To my friends and family who've encouraged me to follow my dreams.

Nithin Agarwal: To the next generation, with endless possibilities.

Sanjay Ranka: I dedicate this book to my wife, Deepa.

Contents

List of Figures

List of Tables

Authors

Yashaswi Karnati is a computer scientist with expertise in machine learning, computer vision and intelligent transportation systems. Having completed a PhD in Computer Science from the University of Florida, he has embarked on a career that sits at the intersection of academic excellence and industry innovation. He is currently working with NVIDIA Corporation, focusing on the development of large scale foundation models.

Dhruv Mahajan completed his Ph.D. from the Department of Computer & Information Science & Engineering, University of Florida in May 2021. He is currently working on advancing Privacy Preserving Machine Learning techniques at Procter & Gamble.

Tania Banerjee is an Assistant Professor in the Information Science and Technology Department at the University of Houston. Her research interests are in the area of intelligent transportation, and smart healthcare.

Rahul Sengupta is a Ph.D. student at the Computer and Information Science Department at the University of Florida, Gainesville, USA. His research interests include applying Machine Learning models to sequential and time-series data, especially in the field of transportation engineering.

Clay Packard is a principal software and systems engineer at HNTB with a focus in transportation technology. Clay provides technical leadership in systems planning, program and project development, and providing subject matter expertise to transportation agencies.

Ryan Casburn, a traffic engineer with over five years of experience, boasts a lifelong passion for optimizing transportation systems. Fascinated by the dynamic interactions within these systems, he specializes in crafting practical solutions based on real-world behaviors rather than purely theoretical models. His diverse project portfolio spans microsimulation, signal retiming, and transportation planning, software development of user-friendly transportation analysis tools. Ryan's expertise and dedication make him a valuable asset in the realm of traffic engineering and integrated corridor management.

Nithin Agarwal, Ph.D., is a faculty member at the University of Florida Transportation Institute, directs the Florida Technology Transfer (T2) Center and the Transportation Safety Center. With over 15 years' experience in transportation engineering, he develops advanced tools and leads research projects to integrate emerging technologies. Dr. Agarwal teaches transportation courses and creates online training programs, collaborating with AASHTO. He influences industry standards through serving on the panel for the National Cooperative Highway Research Program and serves on the board of professional associations like FACERS and ITS Florida.

Anand Rangarajan is Professor, Dept. of CISE, University of Florida. His research interests are machine learning, computer vision, medical and hyperspectral imaging and the science of consciousness.

Jeremy Dilmore is the Transportation Systems Management and Operation Engineer for the Florida Department of Transportation District 5. He has 19 years of experience with the Department, with 13 of those years in Intelligent Transportation Systems and/or Transportation Systems Management and Operations. In this position he is responsible for leading FDOT District 5's technology efforts including working in the fields of signal timing optimization, managed lanes, simulation modeling, and connected and autonomous vehicles.

Sanjay Ranka is a Distinguished Professor in the Department of Computer Information Science and Engineering at University of Florida. His current research interests are high performance computing and big data science with a focus on applications in CFD, healthcare and transportation. He has co-authored four books, 290+ journal and refereed conference articles. He is a Fellow of the IEEE and AAAS. He is an Associate Editor-in-Chief of the Journal of Parallel and Distributed Computing and an Associate Editor for ACM Computing Surveys, Applied Sciences, Applied Intelligence, IEEE/ACM Transactions on Computational Biology and Bioinformatics.

1 Introduction

Mitigating traffic congestion and improving safety are the important cornerstones of transportation for smart cities. An INRIX study [59] found that in 2017, traffic congestion cost nearly $305 billion and caused Americans to lose 97 hours per person annually in gridlock. This costs the U.S. $87 billion annually in lost time. Many drivers are frustrated due to long (but potentially preventable) delays at intersections. The Governors Highway Safety Administration (GHSA) found that pedestrian deaths have steadily increased with 6590 deaths in 2019 an estimated 60 percent increase over 2009. However, traffic management and control are a big challenge in all major urban areas today, especially growing cities. This is because there are limitations on traffic capacity and throughput in most urban areas. Hence, it becomes essential to utilize the existing road infrastructure and networks as efficiently as possible [148, 19].

Traffic intersections, specifically signalized intersections and traffic controllers are an elementary component of all the road networks and are used to solve the problem of traffic conflicts by implementing a basic version of time division multiplexing [148, 149]. However, this comes with its own set of challenges. The operation of signalized intersections represents a specialized facet of traffic management, demanding distinct expertise and resources. The role of signal operations in improving efficiency is frequently undervalued. Even with sufficient resources, including budget and experienced staff, effectively allocating these resources can pose challenges in the absence of comprehensive knowledge regarding the operational performance of various locations.

Addressing these challenges requires a thorough understanding of traffic patterns throughout the transportation network including intersections, road segments between intersections, and even limited-access facilities that comprise the overall transportation network. Current performance measures typically compare travel-times before and after retiming efforts for validation. However, incorporating additional performance metrics, e.g. arrival on green percentage and travel time reliability measures, will provide greater utility supporting additional decisions on prioritizing which corridor of signals need retiming the most, and handling additional, varying patterns as demand fluctuates across times of the day, days of the week, seasons, and special events. One of the key components of traffic management today is the extensive use of Advanced Traffic Management Systems (ATMS). ATMS systems are good at gathering displaying, and reporting raw data from wide variety of disparate systems and sensors that cannot communicate with each other. The next level of abstraction which enables traffic engineers of today to better manage the road networks is Automatic Traffic Signal Performance Measures (ATSPM) Systems. These draw upon the vast body of recent research on the quantification of the performance of signalized intersections, enabling both better management of signalized intersections and manual optimization of intersections and key corridors. ATSPM utilizes high resolution data from signals and automatically converts the data into performance measures that allow a practitioner to understand the performance of any given intersection and the nature of the problem at said intersection (if any). Furthermore, the availability of high resolution (10 Hz) controller logs opens an even broader range of possibilities. This data, in many cases, is available with small latency (a few minutes) making it amenable to real-time decision making and addressing bottlenecks. However, this plethora of information without proper decision-making tools adds a burden on transportation professionals. This is challenging even for small cities comprising of a few hundred traffic intersections. There is a need for a system that provides corridor-level and city-level information in a succinct and actionable form.

The integration of integrated corridor management (ICM) systems, data, and algorithms is crucial for creating effective traffic management systems and is the focus of this book. We first discuss some general concepts on the traditional approaches to traffic management. We next explore the role of data-driven decision support systems, automation, and machine learning in optimizing traffic flow and improving transportation systems, and finally provide an outline for the rest of the book.

1.1 GENERAL CONCEPTS OF TRAFFIC MANAGEMENT

Traffic management practises are examined in this section, including those for limited access roads and arterial roads, along with the emergence of ICM strategies. The discussion also covers factors contributing to congestion, performance metrics used to evaluate transportation systems, and response strategies.

1.1.1 TRADITIONAL APPROACH TO TRAFFIC MANAGEMENT

Limited access and arterial roads have been managed independently due to their distinct jurisdictions and strategies. These roadways are typically highways or freeways that have limited points of access and egress, with interchanges or exits providing the only means of entering or exiting the road. These types of roads are designed for high-speed travel and are often used for longer distance trips. To manage limited access roadways, the focus is on maintaining a consistent flow of traffic, minimizing congestion, and ensuring safe travel at high speeds.

This can be achieved through strategies such as ITS, ramp metering, and speed limit enforcement. As limited access facilities can run at extremely high volume at full capacity, an incident that blocks a lane can have a significant impact, causing fast-growing congestion queues and presents a secondary crash risk due to the high speed of vehicles. Fast incident detection, response, clearance is a priority focus. Closed television (CCTV) and traffic detection is used to monitor, identify, and validate traffic incidents. Dynamic message signs, highway advisory radio, and advanced traveler information systems including mobile applications are used to inform motorists of incidents ahead so they can take caution and alternate routes. Intelligent transportation systems (ITS) is the term that encompasses these tools and their operation.

Limited-access facilities can span across multiple jurisdictions, thus, the states and federal government are responsibility for funding, constructing, and maintaining these facilities. Arterial roads, on the other hand, are roads that are designed to connect local areas and provide access to local destinations, such as businesses, residential areas, and schools. These roads often have intersections where traffic lights or stop signs regulate traffic flow. To manage arterial roads, the focus is on balancing the needs of motorists, pedestrians, and cyclists, as well as providing efficient access to local destinations. Strategies to manage arterial roads may include optimizing traffic signal timing, improving pedestrian and cycling infrastructure, and implementing traffic calming measures. The responsibility for managing arterial roads is typically under the jurisdiction of a local county or city agency.

1.1.2 INTEGRATED CORRIDOR OPERATIONS

ICM is a transportation management approach that seeks to optimize the performance of a transportation corridor by coordinating and integrating the operations of various transportation modes, including highways, transit, and freight. ICM originated in the 1990s when transportation experts and policymakers recognized the need for a more coordinated and integrated approach to managing transportation systems. One of the early examples of ICM can be found in the San Francisco Bay Area, where the Metropolitan Transportation Commission (MTC) launched the ICM Demonstration Program in 1997. The program aimed to reduce congestion and improve travel time reliability along a 40-mile corridor that included portions of interstates 80 and 680, as well as several local streets and transit routes. The program utilized a range of strategies, including real-time traffic monitoring, traveler information systems, and incident management procedures, to improve the flow of traffic and minimize disruptions.

In the years that followed, other regions across the United States began to adopt ICM approaches, including the Minneapolis-St. Paul region, which launched the MnPASS ICM program in 2005, and the Dallas-Fort Worth region, which launched the Integrated Corridor Management Initiative in 2012, and Florida Department of Transportation, which launched the Regional Integrated Corridor Management System (R-ICMS) in 2019.

Today, ICM continues to evolve and expand, with transportation agencies and policymakers around the world recognizing the potential benefits of a more coordinated and integrated approach to transportation management. ICM strategies are being used to address a range of transportation challenges, including congestion, safety, and environmental sustainability, and are helping to improve the performance of transportation corridors in communities of all sizes.

1.1.3 CONGESTION AND INCIDENTS

Traffic congestion refers to the situation where the volume of traffic on a roadway exceeds the roadway's capacity, causing slower speeds and greater travel times, which contributes to an overall reduction in travel times reliability characterized by greater variability in travel times. Congestion can occur at any time of day, but it is typically most severe during peak travel periods, such as rush hour. Several factors contribute to traffic congestion, including:

- High travel demand: Congestion can occur when there are too many vehicles trying to use a roadway, particularly during peak periods when many people are traveling to and from work or school.
- Limited roadway capacity: Congestion can also occur when a roadway design is insufficient for the volume of traffic using it.
- Incidents and crashes: Congestion can be caused by incidents such as crashes, breakdowns, and construction work that reduce the capacity of a roadway or force vehicles to slow down or stop.
- Poor traffic management: Congestion can be exacerbated by poor traffic management practices, such as poorly timed traffic signals or inadequate signage, that make it difficult for vehicles to flow smoothly through an area.

Congestion can negatively impact productivity, fuel consumption, emissions, and quality of life for people living and traveling in congested areas. To address congestion, transportation planners and policymakers may implement a range of strategies, such as improving public transportation options, encouraging shared mobility such as carpooling and ride-share services, optimizing traffic signal timing, and promoting active transportation modes like walking and cycling.

1.1.4 PERFORMANCE METRICS

Traffic performance measures are metrics used on a transportation system or corridor to evaluate the operational performance, treatment effectiveness, or investment priority. These measures assess the efficiency, safety, and effectiveness of a transportation network, and to identify areas where improvements may be needed. Here are some examples of common traffic performance measures:

- Travel time is a time measure of how long it takes vehicles to travel along a roadway or corridor. It can be used to assess the efficiency of a transportation network and to identify areas where travel times are excessively long.
- Travel speed is a measure of how fast vehicles are traveling along a roadway or corridor. It can be used to assess the efficiency of a transportation network and to identify areas where speeds are too low, which can contribute to congestion.
- Level of service is a measure of how well a roadway or intersection is serving the needs of different user types, i.e. cars, trucks, bicycles, and pedestrians. It takes into account factors including travel time, travel speed, and congestion, and is used to evaluate the effectiveness of roadway design and traffic operations.
- Safety measures include metrics such as the quantity and severity of crashes per vehicle mile traveled. These measures can be used to assess the public safety of a transportation network and to prioritize areas where safety improvements are needed.
- Vehicle miles traveled (VMT) is a measure of the total quantity of miles traveled by all vehicles on a roadway or corridor. It can be used to assess the

overall demand for travel on a transportation network and to identify areas where demand is particularly high.
- Mode share is a measure of the percentage of trips made by different modes of transportation, such as cars, bicycles, and public transit. It can be used to assess the effectiveness of transportation policies and programs that are designed to promote alternative modes of transportation.

By using traffic performance measures, transportation planners and policymakers can identify areas of concern and develop targeted strategies to improve the performance and safety of transportation networks.

1.1.5 RESPONSE AND OPTIMIZATION STRATEGIES

There are a vast landscape of traditional response and optimization strategies attempt to address the congestion, incidents, and inefficiencies and measured by the above performance measures. Many mature strategies focus distinctly on arterials or freeways, due to the separation of jurisdiction and established methods. Most are based heavily on engineering judgement and analysis.

Simulation and signal timing optimization is a long-established practice. Data driven analytics has become more prevalent in recent years. Integrated corridor management is fairly new field incorporating data-driven analytics and decision support in real-time. Machine learning techniques are being explored to solve these problems in research for many existing and new techniques to support operations in a more automated fashion to reduce the operations cost and human error from relying purely on manual data entry.

In this book, we provide an overview of traditional and emerging traffic operations, available datasets, analytics, a set of these algorithms, and arrange them in a system that addresses this need. Our models leverage machine learning methodologies for data collected from a large number of intersections to derive key spatio-temporal traffic patterns in a city and then interactively allows a traffic engineer to focus on key challenges or improvements that can be carried out to alleviate them.

1.1.6 TRAFFIC CONTROL SYSTEMS

Traffic control systems are systems that manage the movement of vehicles, pedestrians, and other modes of transportation along roadways and in other transportation settings. These systems typically include a combination of hardware, software, and communication systems that work together to optimize traffic flow and improve safety. Here are some examples of common traffic control systems:

1. Traffic signals are devices that control the flow of vehicles and pedestrians at intersections. They use red, yellow, and green lights to indicate when drivers and pedestrians should stop, proceed with caution, or proceed safely through an intersection.
2. Variable message signs are electronically changeable signs that display messages to motorists about traffic conditions, construction zones, and other

important information. They provide real-time information about traffic incidents, travel times, and road closures. These are also commonly referred to as changeable message signs or dynamic message signs.

3. Roundabouts: Roundabouts are circular intersections where each approaching ingress road segment requires drivers to yield to other vehicles already in the roundabout before entering. Once inside, vehicles drive round about the intersection until freely exiting to the desired egress road segment. They are designed to improve traffic flow and reduce crash risk.
4. Pedestrian crossings are designated areas where pedestrians can cross a roadway safely. They can include crosswalks, pedestrian signals, and other devices designed to improve pedestrian safety.
5. Speed limit enforcement systems use technology such as radar and cameras to detect and enforce speed limits. This can reduce the number of speeding violations and accidents, improving safety.

Intelligent Transportation Systems (ITS) are a set of technologies and services that use real-time data to optimize transportation systems. ITS can include traffic monitoring systems, traveler information systems, and incident management systems. Overall, these systems are critical for managing the movement of traffic and improving safety on roadways and in other transportation settings. By implementing these systems, transportation planners and policymakers can help ensure that transportation systems are efficient, safe, and accessible to all users.

1.2 DATA DRIVEN DECISION SUPPORT SYSTEMS

Data-driven decision support systems and automation are becoming increasingly useful for ICM. These technologies use real-time data and analytics to support transportation agencies and operators make more informed decisions about managing traffic along a corridor. Here are some examples of data-driven decision support systems and automation used in ICM:

1. Traffic Management Centers (TMCs) are centralized command centers that use real-time data to monitor traffic conditions along a corridor. They can receive data from a variety of sources, such as traffic sensors, cameras, and weather stations, and staff use this information to make decisions about traffic management strategies.
2. Adaptive traffic control systems make real-time adjustments to traffic signal timing based on current traffic conditions. These systems improve traffic and reduce congestion along a corridor.
3. Incident management systems: Incident management systems use real-time data to detect and respond to traffic incidents, such as accidents, road closures, and construction zones. These systems can help reduce the impact of incidents on traffic flow and improve safety for drivers and pedestrians.
4. Traveler information systems provide travelers with real-time information about traffic conditions, travel times, and alternative routes. These systems can help travelers make more informed decisions including adjusting the

start time of their trip, the route they take, and when to change their travel plan. In aggregate, this can alleviate congestion in the transportation network through planning, diversion, and avoidance of congestion by the travelers.

5. Predictive analytics use machine learning algorithms that process historical traffic data to predict future traffic conditions. These systems can help transportation agencies and operators make proactive decisions about traffic management strategies to improve the performance of the transportation system.

Overall, data-driven decision support systems and automation are becoming increasingly important for managing traffic flow along corridors. Transportation agencies and operators can use these technologies to make more informed decisions and take proactive actions to improve the performance and safety of the transportation system.

The following section 1.2.1 discusses the increasing how data analytics and machine learning can be used to optimize traffic flow and improve transportation systems. It outlines several applications of these technologies, including traffic flow prediction, route optimization, incident detection and management, integration with ITS, and support for autonomous vehicles. Section 1.2.2 emphasizes the importance of ICM systems, data, and algorithms to create a cohesive traffic management system. Integration benefits include real-time data sharing, automated decision making, improved incident management, and better traveler information, ultimately leading to optimized traffic flow, enhanced safety, and reduced congestion.

1.2.1 DATA ANALYTICS AND MACHINE LEARNING FOR TRAFFIC OPERATIONS

Data analytics and machine learning can be used to optimize traffic and improve transportation systems. These technologies use real-time and historical data to analyze traffic patterns and forecast future traffic conditions. Here are some examples of how data analytics and machine learning can be used for traffic optimization:

1. Traffic flow prediction: Machine learning algorithms can be trained to predict traffic patterns and congestion based on historical data.
2. Route optimization: Data analytics can be used to analyze traffic patterns and identify the most efficient routes for drivers.
3. Incident detection and management: Machine learning algorithms can be used to detect traffic incidents, including crashes and closures, from real-time data. This can enable faster response to incidents and minimize their impact on traffic flow.
4. Autonomous vehicles: Autonomous vehicles use machine learning algorithms to make decisions about speed, acceleration, and other factors that can impact traffic flow. As more autonomous vehicles are introduced to the roadways, machine learning will become increasingly important for optimizing traffic flow and improving safety.

Data analytics and machine learning provide a variety of ways to support traffic optimization. By leveraging these technologies, transportation agencies and operators can improve traffic flow and transportation performance in ways that were not formerly possible by staff, and allowing the current staffing level to manage a larger transportation system through more efficient operations.

1.2.2 INTEGRATION OF ICM SYSTEMS, DATA, AND ALGORITHMS

Integration of ICM systems, data, and algorithms is crucial for creating a cohesive and effective traffic management system. By integrating ICM systems, transportation agencies and operators can more easily collect and share a variety of real-time traffic information among systems that can be leveraged by algorithms to respond to issues and improve the overall performance of transportation systems. Here are some examples of how integration of ICM systems, data, and algorithms can improve traffic management:

1. Real-time data sharing: By integrating ICM systems, transportation agencies and operators can more easily share real-time data about traffic conditions, incidents, and other factors that can impact traffic flow. This information can be used to optimize traffic management strategies and improve the overall performance of the transportation system.
2. Automated decision making: By integrating ICM algorithms into traffic management systems, transportation agencies and operators can automate decision making processes, such as adjusting traffic signal timing or rerouting traffic in response to an incident. This can help reduce the time it takes to respond to incidents.
3. Improved incident management: By integrating ICM systems, transportation agencies and operators can more quickly and effectively respond to incidents. This can help reduce the impact of incidents on traffic flow and improve safety for drivers and pedestrians.
4. Better traveler information: By integrating ICM systems, transportation agencies and operators can provide more accurate and up-to-date traveler information and even suggest alternative routes and trip plans in response to traffic conditions. Drivers can then make more informed decisions on their travel plans and routes to avoid incidents and congestion along a corridor, resulting in less congestion overall.

Overall, integration of ICM systems, data, and algorithms is crucial for creating a cohesive and effective traffic management system. By leveraging real-time data and automated decision making, transportation agencies and operators can improve the safety and performance of the transportation system. This book explores machine learning concepts centered around ICM system concepts.

1.3 OUTLINE OF THE BOOK

This section unfolds a comprehensive journey through the intricacies of modern transportation management systems, addressing key aspects from traffic signal dynamics to advanced machine learning applications.

Chapter 2 provides an in-depth exploration of the pivotal role of traffic signals within transportation management systems, elucidating their core principles and operational modes. It introduces time-space diagrams as a valuable tool for evaluating signal timings and emphasizes their role in assessing coordination between intersections. Furthermore, the chapter discusses the operational intricacies of limited access facilities, emphasizing the importance of meticulous coordination to optimize traffic flow, ensure safety, and reduce congestion. Advanced traveler management system strategies, including Managed Lanes, are also examined for their potential to enhance traffic flow and provide efficient travel options. Lastly, the chapter underscores the significance of integrated corridor management policies and the role of engineering judgment in ensuring safe and effective traffic management, advocating for the integration of operator review and approval to mitigate risks associated with machine error in Intelligent Corridor Management Systems (ICMS).

Chapter 3 describes an integrated corridor management system (ICMS) that implements the ICM strategies using automated functions powered by data in real-time. While some functions are automated, many functions play a part of a workflow using a transportation operator to review, revise, confirm, and finally activate a strategy. The ICMS will require a robust data management subsystem to collect and manage data to power other subsystems. Other subsystems implement business logic around one or more traffic systems or advanced traveler management strategies. Several interfaces exchange information between operators and other actors in the system throughout the workflow to implement the strategies.

Chapter 4 explores traffic datasets that are at the heart of any advanced transportation management system. Datasets provide situational awareness of current conditions, derivations, and predictions necessary for systems and operators to make decisions. Data is collected and derived by several modalities for several usages, as detailed in the following sections.

Chapter 5 briefly explores the critical areas of clustering, outlier detection, and neural networks within the context of machine learning and its application to traffic data analysis. The chapter delves into various algorithms, methodologies, and applications of these concepts, with a special focus on traffic data analysis, including traffic state prediction, intersection performance, and detection of traffic interruptions. By giving a brief introduction to machine learning techniques, this chapter aims to provide insights into how these can be leveraged for applications in data driven management of traffic networks.

Chapter 6 discusses the role of traffic simulation frameworks. The different types of simulation frameworks are discussed, based on the complexity of modeling. Some important traffic simulation frameworks, that are often used in conjunction with Machine Learning algorithms, either as sources of training data, or as environments of Reinforcement Learning algorithms have been presented. How a traffic simulation can be calibrated based on real-world considerations is treated in detail in this chapter.

Chapter 7 describes intersection detector diagnostics. Current traffic signal controllers are capable of logging events (signal events, vehicle arrival, and departure) at very high resolutions (usually 10 Hz). The high resolution data rates enable the computation and study of various (granular) measures of effectiveness. However, without knowing the location of specific detectors on an intersection and the phases they are mapped to, a number of measures of effectiveness (of signal performance) cannot be evaluated. These mappings may not be available or up to date for many practical reasons (e.g., old infrastructure, mappings not machine readable, maintenance, addition of new lanes, etc.). An inference engine is presented to map detectors to phases and distinguish between the stop bar and advance detectors or, in other words, infer the location of the vehicle detectors with reference to the intersection.

Chapter 8 delves into intersection performance. Traffic signals are installed at road intersections to control the flow of traffic. An optimally operating traffic signal improves the efficiency of traffic flow while maintaining safety. The effectiveness of traffic signals has a significant impact on travel time for vehicular traffic. There are several measures of effectiveness (MoE) for traffic signals. A work-flow to automatically score and rank the intersections is presented in a region based on their performance, and group the intersections that show similar behavior, thereby highlighting patterns of similarity. The process, also helps detect potential bottlenecks in the region of interest.

Chapter 9 describes machine learning approaches for detection of traffic interruptions which is a critical aspect of managing traffic on urban road networks. This chapter outlines a semi-supervised strategy to automatically detect traffic interruptions occurring on arteries using high resolution data from widely deployed inductive loop detectors. The techniques highlighted in this paper are tested on data collected from detectors installed on more than 300 signalized intersections over a 6 month period, hence can detect interruptions with high precision and recall.

Chapter 10 discusses an algorithm for estimating turning movement counts. Predicting intersection turning movements is an important task for urban traffic analysis, planning, and signal control. However, traffic flow dynamics in the vicinity of urban arterial intersections is a complex and nonlinear phenomenon influenced by factors such as signal timing plan, road geometry, driver behaviors, queuing, etc. Deep neural networks that are capable of directly learning an abstract representation of intersection traffic dynamics using detector actuation waveforms and signal state information are presented. The chapter describes in detail, neural network models for predict turning movement counts with greater accuracy when higher resolution data are provided.

Chapter 11 presents an end to end solution for automatically generating intersection coordination plans. The road network infrastructure (signal controllers and detectors) continuously generates data that can be transformed and used to evaluate the performance of signalized intersections. Current systems that focus on automatically converting the raw data into measures of effectiveness have proven extremely useful in alleviating intersection performance issues. However, these systems are not well suited for automatically generating recommendations or suggesting fixes as needed. The use of machine learning and data compression techniques to build a recommendation system is demonstrated in this chapter.

Chapter 12 explores the intricate dynamics of traffic flow near urban arterial intersections, a multifaceted and nonlinear phenomenon shaped by various factors including signal timing plans, road geometry, and driver behaviors. While predicting such dynamics is crucial for urban traffic signal control and planning, current methods rely on microscopic simulation, which, despite its utility, faces limitations in incorporating high-resolution loop detector data provided by automated traffic signal performance measures (ATSPM) systems. To address this gap, the chapter introduces neural network models capable of harnessing waveform data collected from multiple sensors to effectively model traffic dynamics both within and across intersections. These models offer insights into platoon dispersion and the progression of vehicles across corridors under different signal timing plans, potentially informing signal timing optimization software. Notably, these methods offer significant computational efficiency, being three to four orders of magnitude faster than microscopic simulations.

Chapter 13 deals with modeling measures of effectiveness for intersection performance. Microscopic simulation-based approaches are extensively used for determining good signal timing plans on traffic intersections. Measures of Effectiveness (MoEs) such as wait time, throughput, fuel consumption, emission, and delays can be derived for variable signal timing parameters, traffic flow patterns, etc. However, these techniques are computationally intensive, especially when the number of signal timing scenarios to be simulated are large. This chapter presents InterTwin, a Deep Neural Network architecture based on Spatial Graph Convolution and Encoder-Decoder Recurrent networks that can predict the MoEs efficiently and accurately for a wide variety of signal timing and traffic patterns. These methods can generate probability distributions of MoEs and are not limited to mean and standard deviation. Additionally, GPU implementations using InterTwin can derive MoEs, at least four to five orders of magnitude faster than microscopic simulations on a conventional 32 core CPU machine.

Chapter 14 presents the field of computational signal timing optimization, where various mathematical modeling approaches, as well as machine learning approaches are discussed. The chapter also discusses how signal offset optimization may be performed using a calibrated simulation model.

Finally, Chapter 15 presents some techniques and tools for traffic data visualization. The advent of new traffic data collection tools such as high-resolution signalized intersection controller logs opens up a new space of possibilities for traffic management. This chapter presents visualization tools to provide traffic engineers with suitable interfaces, thereby enabling new insights into traffic signal performance management. The eventual goal of this chapter is to enable automated analysis and help create operational performance measures for signalized intersections while aiding traffic administrators in their quest to design 21st century signal policies.

2 Traffic Engineering and Operations Background

Abstract: This chapter delves into the critical role of traffic signals within transportation management systems, elucidating their fundamentals and various operational modes. It introduces the concept of time-space diagrams for evaluating signal timings, emphasizing their utility in assessing coordination between intersections. Moreover, it distinguishes between signal operational modes, highlighting the significance of stages and phases in signal timing.

Additionally, the chapter discusses limited access facility operations, focusing on the management of highways and transportation facilities with restricted access points. Key aspects such as traffic management, incident response, toll collection, intelligent transportation systems (ITS), and maintenance are outlined, emphasizing the need for meticulous coordination to optimize traffic flow, ensure safety, and reduce congestion along limited access facilities.

Furthermore, advanced traveler management system strategies are explored, including Managed Lanes, which are designed to enhance traffic flow and provide additional travel options for drivers. The section emphasizes the benefits of managed lanes in improving safety, reducing congestion, and providing more efficient travel options.

Lastly, the chapter addresses integrated corridor management policies and constraints, highlighting the paramount importance of public safety in traffic engineering and operations. It underscores the role of engineering judgment and oversight in ensuring safe and effective traffic management, advocating for the integration of operator review and approval in Intelligent Corridor Management System (ICMS) recommendations to mitigate risks associated with machine error.

2.1 SIGNAL TIMING FUNDAMENTALS

Traffic signals are a critical component of a transportation management system. This section explores the fundamentals of traffic signals through which all integrated corridor management must work within the bounds of different traffic signal modes.

2.1.1 SPACE TIME DIAGRAM

When evaluating and creating traffic signal timings, space time diagrams are often used to estimate their performance. Time space diagrams are useful for evaluating the coordination between intersections. On these diagrams, distance along the corridor is one axis (usually horizontal) and time is the other axis (usually vertical). Time that a signal is green and red are shown along the time axis for each intersection. A diagonal line can then be drawn between signals with a slope based on the corridor

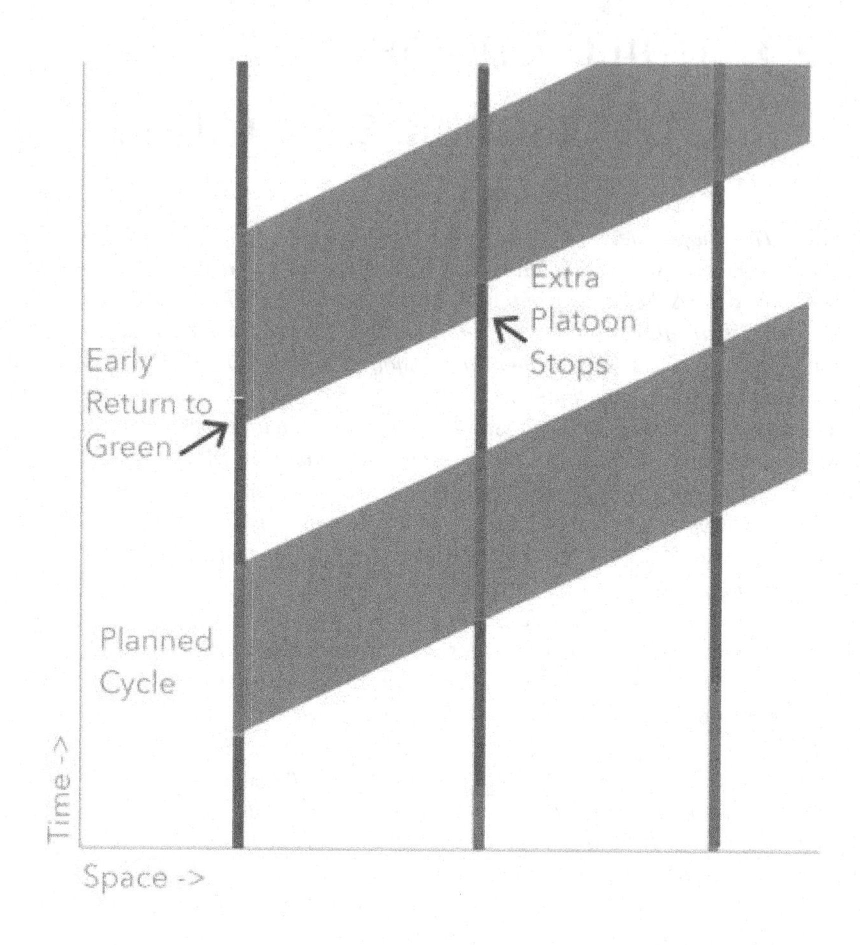

Figure 2.1: Space and Time Diagram.

travel speed. Based on these lines, generally the largest gap of time that will allow travel through all intersections is defined as the green band.

There are some key considerations that must be made when evaluating these time space diagrams. These considerations are described as follows:

Queue Discharge: If a platoon is shown to arrive near or at the start of green of an intersection, any queue that may be on the approach should be considered. Ideally, the platoon should arrive following the discharge of the queue; otherwise, there is an increased risk of collision with a platoon arriving to a green light which still has stopped traffic. There is also a decreased efficiency with platoon vehicles needing to slow or stop prior to the queue clearing.

Platoon Dispersion: Platoon dispersion is the tendency of a group of vehicles to spread out over time. Platoon dispersion is caused by a variety of factors including driver behavior tending to spread out further from intersections, access point turns, and random differences in driver speed preferences. Each corridor will likely have different platoon dispersion patterns.

The saturation flow rate (the maximum rate of departure from an intersection) of a movement may be 1800 vehicles per hour per lane. However, at a downstream intersection, the arrival flow rate may only be 1600 vehicles per hour per lane (for a longer duration). Platoons generally continue to disperse the longer they continue through the corridor, except if they have reasons to bunch back up (for example arriving at an intersection before a queue has cleared).

If the suggestion in 1 above is taken to have platoons arrive following the queue clearance, there may end up being multiple platoons with their own dispersions within the green band. The queue discharge begins as a new platoon which disperses over time, then the arriving vehicles are their own platoon which continues to disperse as it continues down the corridor. This effectively means that the green band needs to continue to get wider along the corridor's length to serve the full platoons. However, generally, due to restrictions from other movements, the platoon ends up being cut off at particular intersections such that the extreme bounds of the platoon receive a red light and end up queuing at that intersection. This can also be achieved through strategic platoon arrival with an existing queue to cause the platoon to tighten back to close to saturation flow rate.

Corridor Travel Patterns and Purpose: Every corridor has different travel patterns and different purposes. Travel patterns can be explored through the use of turning movement counts and origin-destination data. Some intersections likely have significant rates of turning movements. In these cases, it may make sense to have coordination for these turning movements and less coordination on the through movement. This may show up on a time-space diagram as a severe restriction in green band, but the additional context of the corridor travel patterns may make that the most optimized signal timing.

Corridor purpose must also be explored. In Florida, corridor purpose is often expressed through the combination of context classification and access classification. Other states have similar methods for expressing roadway purpose. Context classification is related to the area type and surrounding land uses. Access classification is related to the number of cross streets, driveways, and median openings on the roadway. It is important to recognize the differences in corridor priorities when evaluating signal timing and time-space diagrams.

For example, a corridor in a suburban residential area, with few access points and median openings, but dense enough signal spacing for coordination is likely serving primarily through traffic. The primary travel patterns would be vehicles leaving subdivisions in the morning, then traveling through the corridor to an employment center, with the opposite in the afternoon. In this case, optimizing for a large green band to serve the through traffic is ideal. Vehicles turning out of subdivisions may come to the next signal being red, but following that they could have only green lights for the rest of their trip.

Another example is a corridor in a suburban commercial area, with a large number of access points, a continuous left turn lane median, and dense signal spacing. In this case, travel patterns are much less likely to be through, but rather be destined to or coming from one of the many access points. The high number of access points increases platoon dispersion. In this case, having a large green band with low corridor travel time is likely much less beneficial than having a smaller through green band, while adequately serving all turning demand. In this case, vehicles going through the corridor may be stopped at a few red lights, while vehicles leaving access points or side streets are able to have a lower travel time.

2.1.2 TRAFFIC SIGNAL MODES

Traffic signals can act in several different modes. Each mode type has different operating characteristics and ideal operating environments. Some signal operational modes use stages and other operational modes use phases. A stage represents the status of all signal indications at the intersection. A phase represents the status of a single movement at the intersection. For example, a simple signal mode may first be green in the east and west directions and red in the north and south directions. Then after some time (and a yellow indication), the east and west directions are red, and the north and south directions signal indications are green. This would be an example of a two-stage intersection. The same intersection could be represented as a four-phase signal (one phase per direction). Phases become more important at more complicated phases which may have the possibility for a phase to end in one direction before the same phase ends in the opposite direction.

Phases are best represented utilizing a ring-barrier diagram, as shown in Figure 2.2. A ring-barrier diagram represents each of the phases of an intersection and demonstrates which phases are able to end early.

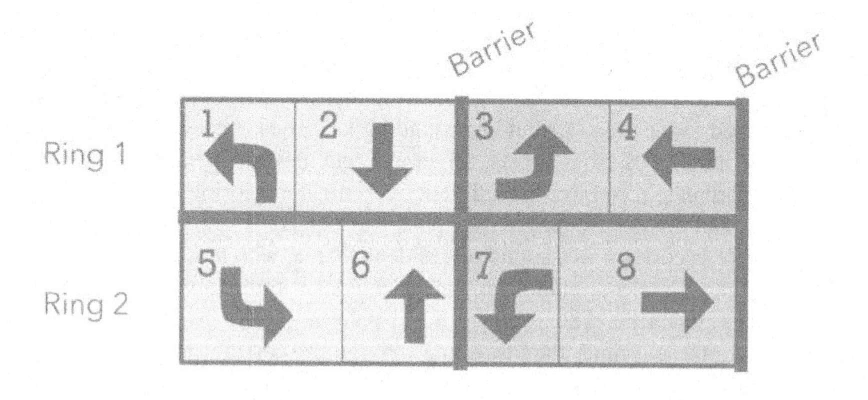

Figure 2.2: Ring and Barrier Diagram.

Typically, the major street through movements are phases 2 and 6, the major street left turn movements are phases 1 and 5. The minor street through movements are phases 4 and 8 and the minor street left turn movements are phases 3 and 7. The rings are the rows of the diagram. The diagram also has two barriers. One barrier occurs after all major street phases, the other barrier occurs after all minor street phases. The importance of the barriers is that both rings must cross the barrier at the same time. Phases within a ring cannot operate at the same time and phases on opposite sides of a barrier cannot operate at the same time. However, phases in different rings on the same side of the barrier can operate at the same time. For example, in the diagram above, phase 5 may end early and phase 6 can start early, even if phase 1 is still running. Also, phase 4 may be ready to end early, but it cannot end until phase 8 is also ready to end so that the barrier is crossed at the same time.

The different possible modes for a traffic signal operations are described in the following subsections.

Fixed time mode

Fixed time mode is the simplest mode of signal operations. In this mode, the signal does not respond to traffic conditions and simply repeats a consistent cycle of signal stages.

Free Mode

Free mode on a traffic signal has the highest flexibility to respond to traffic, but generally does not work well with nearby intersections. Each phase generally has at least three timers.

- Minimum Timer – For safety and to match driver expectations, each phase must be green for at least a certain amount time before yellow can begin. This timer begins at the start of green and counts down continuously to 0. Phase termination is inhibited before the timer is expired. At some intersections, detection is not placed at the stop bar. In these cases, a phase is called using an upstream approach detector. In order to appropriately serve all vehicles which have arrived (since there is no indication of vehicles still present at the stop bar in a queue), the signal can be set up to add green time for each vehicle that is observed on the approach during yellow and red.
- Maximum Timer – To best serve traffic, there is a maximum amount of time a phase can run while there is demand on an opposing phase. This timer starts to run when demand on an opposing phase is detected. Phase termination is forced when the timer expires.
- Vehicle Extension Timer – This timer is how a signal determines when a phase can end early. This timer starts at the beginning of the phase and it resets each time a vehicle is detected on the approach for that phase. If the timer expires, the phase is eligible to end. For example, if the starting value for this timer is three seconds, the phase will be eligible to end early if the headway on the approach increases beyond three seconds.

The vehicle extension timer has several ways to adjust how it works. One set of parameters is called volume-density mode. In this mode, after a set period of time the value this timer is reset to begins to reduce for a period of time. This enables the intersection to gap out sooner to better balance multiple approaches with high volumes.

Another parameter controls whether the vehicle extension timer can be reset after it has expired. This is called simultaneous gap out. If a phase with this setting has an expired vehicle extension timer but is unable to gap out either because minimum green is not met or the phase is at the barrier and must wait for the second phase to finish, then the vehicle extension timer would be reset if another vehicle is detected on that approach. In this way, phases which are required to end together will only end early if both phases simultaneously have expired vehicle extension timers. If simultaneous gap out is not set, then both phases only have to both expire their vehicle extension timer, not necessarily at the same time.

A free-running signal can also have different recall modes for each phase.

- Minimum Recall – This mimics a call being placed on the phase as soon as red starts which clears as soon as green starts. This guarantees the phase will run.
- Maximum Recall – This mimics a call being placed continuously. The phase will run for as long as possible. An intersection running with all phases on maximum recall is identical to a fixed time signal.
- Pedestrian Recall – This mimics a continuous pedestrian call being placed on a corresponding phase.

Coordination Mode

Coordinated mode is very similar to Free Mode with additional features to allow for nearby intersections to be coordinated together. For this mode a coordination point is set. There are several different places a coordination point can be set, but often it is at one of the barriers (generally the barrier after the major street phases). Additionally, cycle length and offset values are set. The cycle length defines how frequently the coordination point will be hit. The offset defines an offset from a master clock for when the coordination point will be hit. For example, if the offset for an intersection is 20, the coordination point (also known as the local clock 0) will occur when the master clock is at 20.

Historically, cycle lengths were defined by measuring grid power frequency and offsets were set by hand using watches. Modern controllers, generally utilize the network time protocol to synchronize clocks and define the offset from midnight of the current day. With this set up, even if other phases ended early, the coordinated phases (generally the major street) would hold in green until the coordination point occurs.

Each phase in a coordinated signal also has a split length, which is a duration that defines an additional timer on the phase. This additional timer is the force-off timer which keeps the time in the cycle when the phase must end in order to serve all other phases split times. A phase may still have a max timer running (in which case the phase would end whenever the max time or force off time is met), otherwise (and in most cases) the max timer is inhibited, or otherwise ignored.

There are two modes which are critical to the operation of a coordinated signal. These define how the force off point is defined. A force-off point at a coordinated intersection is the point where a phase must end to be able to serve the rest of the intersection phases while maintaining coordination.

- Fixed force-off mode – In fixed force-off mode, the force-off points do not move. In this way, if an earlier phase ends early, the next phase has the opportunity to use that extra time.
- Floating force-off mode – In floating force-off mode, the distance between the force-off points stay the same. So if one phase ends 10 seconds earlier, the force-off point for the next phases moves 10 seconds earlier. The coordination point always acts as a fixed point. In this way, any extra time from all phases is shifted to the coordinated phase.

Coordinated signals typically have time-of-day schedules to adjust their operating conditions based on the prevalent travel pattern. Each time of day pattern generally has a unique cycle length, offset, and split times. This allows a signal to be flexible to traffic conditions throughout the day.

When time of day patterns switch (as well as in some other conditions), the coordination point may change suddenly. If the signal instantly recognized this new coordination point, some phases may have unrealistic times or be skipped completely. Instead, the signal enters into a transition mode. There are two main methods for transition:

1. Long-way transition increases the phase lengths to change offset.
2. Short-way transition decreases phase lengths to change offsets.

Each transition method has a maximum difference that can be set on each phase (generally defined as a percentage) and must continue to serve the minimum times defined for each phase. A transition is often set to occur in the best way as determined by the signal controller. While in transition, the signal adds time or removes time from phases to gradually ease into the new coordination point. The other parameters of the phase must still be met (minimum green, yellow, all red, etc).

2.1.3 SIGNAL COORDINATION CHALLENGES

In many cases, it is not possible to perfectly coordinate a system. Variability in traffic volumes, vehicle speed, and interruptions can lead to poor coordination. This section elaborates on these challenges

Platoon Dispersion

Platoon dispersion is one cause for poor coordination. Platoon dispersion can occur where there is a large distance between signals and/or a large number of mid-block turning movements. A platoon leaves a traffic signal at a high density and at a flow rate close to saturation flow rate. Over the distance between signals this grouping of vehicles spreads out to a lower density and lower instantaneous flow rate (even if the hourly flow rate is identical). This spreading of the queue is natural as vehicles increase gaps at higher speeds after leaving the intersection. Turning movements with mid-block intersections and driveways can also slow portions of the platoon more than others increasing the dispersion of the platoon. Interruptions occurring between signals can also cause platoons to disperse.

Platoon dispersion can lead to poor coordination as to serve the entire dispersed platoon, the green time for the coordinated movement would need to be too long. As a platoon can continue to disperse as it travels through multiple green signals, this can become a problem even with short signal spacing. At some point, it may be most efficient to purposely stop the platoon to have it bunch up to a higher density. It may also make sense to have the front portion of the platoon have to stop at each intersection for a small amount of time to bunch them back with the rest of the platoon and also connect them with the queue departure platoon while also stopping the green before the full platoon has been served to prevent the tail of the platoon from spreading out too much.

Early Return To Green

Early Return to Green is a situation that can lead to poor coordination and system performance. In many cases it is desirable to program signal timing so that all movements can serve all anticipated demand for every cycle. It may also be desired to program all phases to cover the entire duration of the pedestrian timing, even if the pedestrian signal is not on recall and pedestrian movements are rare. These conditions can lead to early return to green. Early return to green is when the coordinated movement starts green sooner than scheduled by coordination, caused by other phases in the cycle gapping out. Early return to green can cause the platoon departing from this signal to reach the downstream signal before it turns green. A coordinated system that has the coordinated movements showing up at a perfect time when all phases max out will look like a nearly uncoordinated system with the coordinated movements arriving on red randomly when phases gap out early. Floating force-off (as opposed to fixed force-off) makes this even worse.

It is just as possible that the coordinated movements end up in a demand starvation condition (when the light is green but there is no demand) because that signal returned to green early, but the upstream signal did not return to green early.

In many cases, the impact of early return to green is not particularly obvious, particularly during the peak hours as signals are not coordinated in a way that they are in the best coordination during max outs, but are in the best coordination with regular traffic conditions, meaning regular early returns to green are taken into account in the offset determination. However, this can make the coordinated system somewhat unstable as a change in traffic patterns could mean the offsets need to change.

In some cases, it may make sense for coordinated systems to use fixed force off along with phase timings that lead to the final phase before the coordinated phase to force off (as opposed to gap out) in the vast majority of cycles. One should also consider the phase order where possible to move those phases most likely to gap out to occur earlier in the cycle (such that there are more phases after the gapping out phase before the coordinated phase starts).

As already mentioned, including all required pedestrian time in a side street phase split time can also lead to early return to green if the pedestrian is not on recall, pedestrians are rare, and the pedestrian crossing time is much higher than the time required for vehicles typically. Many agencies avoid setting phase times shorter than the time required (possible using the "stop-in-walk" controller setting, or similar) for pedestrians because doing so forces the controller to go into a transition period each time there is a pedestrian.

However, over sizing a vehicle phase time to accommodate an occasional pedestrian can lead to very similar impacts to transition through the early return to green affect. In this case, the signal returns to green early every cycle, except for when there is a pedestrian, leading coordination to be based on this "false" green return time rather than the actual planned coordination time. This can make coordination challenging.

In some low volume pedestrian environments, it may be best to plan cycle lengths and phase lengths first on the vehicle volumes only. With strategic (i.e., not uniform) increases to phase lengths and adjustments to transition rates such that any phase overage leading to transition from pedestrians could be recovered within one cycle using short-way transition. Using this method would likely lead to lower cycle lengths compared to always accommodating all pedestrian movements. Lower cycle lengths are better for vehicle delays, queue lengths, and pedestrian delay. This method is also likely going to lead to improved coordination with reduced early returns to green, which would lead to improved system performance without pedestrians. When pedestrians do arrive, the coordinated movements would start later (essentially the same effect as the other alternative with regular early return to green). In the side streets the next cycle would likely end up being somewhat shorter, which may be fine given they ran longer the previous cycle. If timed as described, the coordinated movement would start on schedule the next cycle.

Two Way Travel

Two Way travel can also lead to coordination challenges. In a one way fixed time coordinated system, the difference in offset between two signals would be approximately equal to the typical travel time between the signals. In a two-way system, perfect coordination in one direction would lead to one offset recommendation and coordination in the opposite direction would lead to a different offset recommendation. These can be significantly different, and varying signal spacing can also lead to challenges as well.

There is often no best way to handle this. In a two-way system with high directionality, such as during rush hour, it is common to time signals based on a single direction only. When traffic patterns are not highly directional, sometimes high cycle lengths are used such that the platoon arrivals for both directions occur during green, but this can lead to an under utilized coordinated duration with higher delays and queues on the side streets compared to a lower cycle length. With supporting signal hardware, phase order may be adjusted so that one direction has a protected green before the through movement (lead phasing) and the other direction has a protected green after the through movement (lag phasing). In this way, the cycle length can be made shorter while still accommodating the different platoon arrival times.

Adaptive Signal Control

There are various types and brands of adaptive signal control systems. In general, adaptive signal control aims to optimize signal timing in real time based on observed traffic flow conditions. These types of systems are generally best utilized on corridors with varying traffic patterns. Corridors with very regular traffic patterns can often better be served with high quality time-of-day plans.

There are various methods of performing adaptive signal control. Some systems utilize an external control box which sends the main signal controller commands via NTCIP and may interrupt detection information and forward to the controller different detection data to help the controller do what the adaptive signal control system thinks is best.

Other adaptive signal control systems are integrated into the signal controller itself. These adaptive signal control systems still utilize the main signal controller logic, but can dynamically set phase split times, cycle lengths, and offsets to better match the observed traffic.

2.1.4 SIGNAL PRIORITY AND PREEMPTION

To support the priority throughput of certain vehicles for the overall benefit of the transportation system and to support the time-critical arrival of first responders to a scene, many traffic signal are instrumented with a signal priority and preemption module. Preemption will specifically interrupt the phase of a traffic signal to allow the priority vehicle through the intersection. Priority will make more subtle adjustments to the phase durations in attempt to align the active phase for the approaching vehicle being given priority. This can impact the performance of the roadway for a short duration, especially for traffic signals operating in coordination.

2.2 LIMITED ACCESS FACILITY OPERATIONS

Limited Access Facility Operations refers to the management and operation of highways and other transportation facilities that have limited access points. These facilities are designed to provide high-speed and efficient travel for long-distance trips, and they typically have few access points or interchanges. Here are some key aspects of limited access facility operations:

1. Traffic management: Limited access facilities require careful management of traffic flow to ensure safety and efficiency. This includes monitoring traffic conditions, adjusting speed limits, and implementing dynamic lane controls.
2. Incident management: Limited access facilities are prone to incidents, e.g. crashes and disabled vehicles, which can cause significant disruption to traffic flow. Incident management strategies include quick response times, efficient clearance of the incident scene, and effective communication with drivers.
3. Toll collection: Many limited access facilities are toll roads, which require efficient and effective toll collection systems. This includes automated toll collection using electronic toll collection (ETC) systems, as well as manual toll collection at toll booths.
4. Maintenance and construction: Limited access facilities require regular maintenance and occasional construction or expansion. This work must be carefully coordinated to minimize disruption to traffic flow.

Overall, limited access facility operations require careful management and coordination of traffic flow, incident response, toll collection, and maintenance activities.

2.2.1 TRAFFIC INCIDENT AND RESPONSE

Limited access traffic incidents include accidents, disabled vehicles, or other events that disrupt traffic flow on highways or other limited access facilities. Due to the high speed and limited access points of these facilities, incidents can quickly lead to congestion and safety risks for drivers particularly including secondary rear-end crashes. Thus, for limited access facilities, responding to incidents is a priority activity in traffic management. Response activities are characterized and include the following aspects:

1. Quick response times: Rapid response times are critical for mitigating the impact of incidents on traffic flow. Delays in response increase the impact of the initial incident and the likelihood of a secondary incident occurring. Transportation agencies and operators typically have incident management teams that are trained to respond quickly to incidents and clear the scene as efficiently as possible.
2. Traffic control: Incident response teams use traffic control devices such as cones, signs, and electronic message boards to manage traffic flow around the incident scene. The Manual of Uniform Traffic Control Devices includes standardized temporary maintenance of traffic signs and procedures. This helps to reduce congestion and prevent secondary accidents during the incident.
3. Incident clearance: Once the incident scene has been stabilized, the focus shifts to clearing the scene as quickly and safely as possible. This may

involve towing vehicles, removing debris from the roadway, and addressing any environmental hazards such as spilled fuel or other hazardous materials.

4. Communication with drivers: Effective communication with drivers is critical for reducing confusion and frustration around incident scenes. Transportation agencies and operators use electronic message boards, social media, and other channels to provide real-time updates on traffic conditions and alternate routes.
5. Incident analysis: After an incident has been cleared, transportation agencies and operators may conduct an analysis to determine the cause of the incident and identify any opportunities for improving incident response in the future.

By using these strategies to efficiently respond and resolve traffic incidents, their impact on safety and traffic flow can be minimized.

2.2.2 SPECIAL OPERATING MODES: NORMAL, EVACUATION, SHOULD USE, PRIORITY/PREEMPTION

Limited access facilities support several special operating modes that need to be considered. During a large-scale event such as an imminent hurricane, the facility may be designated as an evacuation route. There will be a significantly larger volume in one direction during the evacuation. The facility will be extremely sensitive to any possible capacity reduction events. Off-ramp backups from vehicles seeking fuel, vehicles running out of fuel due to high demand and less availability are a few possibilities. Shoulder use is a special operating mode where the shoulders are designated as mainline travel lanes. Shoulder use increases the capacity of the freeway, but impacts the ability to pull over or handle a traffic incident, especially if the shoulder is adjacent to a guardrail or on a bridge.

Priority mode is normally temporary to allow a first responder or official vehicle to travel faster than the general-purpose vehicle flow. Vehicles will change lanes usually to the slower lane until the priority vehicle passes.

ICM should take these modes into consideration as a constraint or as a strategy. During an evacuation, adjacent arterials may also have an increased demand to provide additional capacity for the evacuation, and should be optimized accordingly to allow additional green time for the higher demand in the movements in the direction of the evacuation. ICM could utilize shoulder use as a strategy to mitigate severe congestion or a response to a capacity restricting event such as a crash.

2.2.3 TRAFFIC CONTROL SYSTEMS

Limited access facilities employ a variety of ITS devices to detect information from motorists and disseminate information to drivers. Advanced systems and strategies are covered in the following section, while several common ITS devices are covered in the list below:

1. Advanced Transportation Management System (ATMS) is a central software platform that integrates the traffic control systems and devices to facilitate

traffic operations. An ATMS collects and manages data from data generating devices and data feeds, and provides a control system with a user interface to the operator to manage the devices, data, and implement management strategies. The ATMS also archives data for later analysis, performance measures reporting, training and demonstration.

2. Vehicle detection systems provide continuous, real-time traffic conditions and are discussed further in the traffic datasets section.
3. Traffic cameras provide live video of traffic. Operators can use live traffic video to visually verify incidents and monitor traffic in detail. Video can also be used by machine learning and AI to automate the detection of incidents, and to extract additional traffic conditions and further analytics could be used to develop origin destination, traffic and freight demand models, and even identification and vehicle of interest acquisition.
4. Dynamic message signs (DMS) are electronic roadside signs that can be changed by the ATMS. DMS provide brief, visual messages. DMS formerly had a black background with amber text, but now can have full color matrix to present images such as roadway shields and warning icons recognizable from static signs that follow the Manual on Uniform Traffic Control Devices (MUTCD). DMS are used to present travel times, incident ahead warnings, and public safety message campaigns.
5. Highway advisory radio (HAR) systems are installed on the roadway and broadcast audio messages to motorists. These systems have a range of up to several miles, and can provide quite a bit more content than a DMS sign can or than is safe to include in a connected vehicle message. The message can be more verbose to overcome the limitation that messages cannot be targeted to vehicles traveling in a certain direction. HAR systems can include a static sign with beacons indicating to motorists when an alert is included in the broadcast to encourage motorists to tune their radios into the HAR station's frequency to listen to the broadcast.
6. Connected Vehicle Traveler Information Messages (TIM) and Roadside Safety Alerts (RSA) are additional messaging tools to send messages to motorists within their vehicles. While DMS are positioned at a fixed location, TIMs and RSAs can be broadcasted to vehicles and presented to motorists within a specific location and heading, providing much more flexibility as to the locale of where the vehicle is when the message is presented.
7. Roadway Weather Information Systems (RWIS) detect several weather measurements such as temperature, wind speed, pavement conditions, precipitation at strategic locations along the roadway and can be used to detect a weather hazard for notifying motorists.
8. Wrong way driving (WWD) detection systems deployed at exit ramps can be useful in incident detection of the WWD incident type.

2.3 ADVANCED TRAVELER MANAGEMENT STRATEGIES

2.3.1 MANAGED LANES

Managed lanes, also known as high-occupancy toll (HOT) lanes or express lanes, are a type of roadway facility designed to improve traffic flow and increase travel options for drivers. Managed lanes typically involve converting existing lanes on a highway or building new lanes that are separated from the general-purpose lanes by physical barriers.

Managed lanes can improve safety by reducing congestion and reducing the likelihood of crashes. By separating high-speed traffic from the general-purpose lanes, managed lanes can help prevent bottlenecks and reduce the likelihood of collisions.

Managed lanes are an effective strategy for managing traffic flow and increasing travel options for drivers. By providing a faster and more reliable travel option for those willing to pay a toll, managed lanes can reduce congestion, improve safety, and help drivers reach their destinations more efficiently.

Managed lanes are typically tolled, meaning drivers must pay a fee to use the lanes. The tolls may be set based on current traffic conditions and can vary depending on the time of day, level of congestion, or other factors. Managed lanes may have access restrictions, such as requiring a minimum number of occupants in a vehicle or limiting access to certain types of vehicles, such as buses or motorcycles, to use the lanes for free or even to use the lanes exclusively.

2.3.2 RAMP METERING

Ramp metering involves controlling the release rate at which vehicles pass through the on-ramp to enter the mainline traffic flow of a limited-access facility, i.e. highway or other high-speed roadway. Ramp metering can reducing occupancy spikes, and thereby reduce congestion and reduce the likelihood of crashes that could be attributed to vehicles entering the mainline when it is already at or near full occupancy.

Ramp metering systems use traffic signals at the on-ramp to indicate when vehicles may enter the mainline traffic. The signals may be activated manually or by automated systems that adjust the metering rate based on the speed, volume, and occupancy of the mainline.

Ramp metering systems rely on real-time monitoring of traffic conditions to determine the optimal timing for ramp signals. This may involve sensors embedded in the road, cameras, LiDAR, etc. The determination can be done dynamically by the system, or using operational judgement of a traffic management center operator in real-time. Alternatively, an algorithm or a system can use analysis of historical data to develop time of day ramp metering rate tables.

2.3.3 DYNAMIC LANE CONTROL

Dynamic lane control is a traffic management strategy that involves adjusting the use of lanes on highways and other roads by dynamically indicating lane closures in order to optimize traffic flow and increase safety. Dynamic lane control systems use lane control signs, oftentimes a large, illuminated arrow, to change the configuration of lanes as being open and normally operating, open but experiencing congestion, or closed. These adjustments are made based on current traffic conditions in real-time. By using lane closures and other lane configurations to divert traffic around incidents, dynamic lane control systems can help prevent secondary accidents and reduce delays.

Dynamic lane control requires variable lane markings that can be changed electronically to indicate the current use of each lane. This may involve opening or closing lanes, changing the direction of travel in a lane, or allowing use of a shoulder lane during peak travel periods. Implementing dynamic lane control systems can be challenging due to the need for real-time monitoring, accurate traffic modeling, and effective communication with drivers. In addition, dynamic lane control systems require significant investment in technology and infrastructure.

2.3.4 VARIABLE SPEED LIMIT

Variable speed limits (VSL) are a traffic management strategy used to optimize traffic flow and increase safety on highways and other high-speed roads. Instead of using a fixed speed limit, VSL systems adjust speed limits in real-time based on current traffic conditions, weather, and other factors.

VSL systems can improve safety by reducing the speed differential between vehicles, which can reduce the likelihood of accidents. VSL systems can also be used to slow traffic during adverse weather conditions or other hazardous events.

VSL utilizes real-time monitoring of traffic conditions to determine the appropriate speed limit for a given roadway segment. This may involve sensors that are embedded in the road, cameras, etc.

VSL systems also use traffic modeling to predict how traffic flow will change based on different speed limits. By simulating different scenarios, VSL systems can determine the optimal speed limit to reduce congestion and improve safety.

Effective communication with drivers is critical for the success of VSL systems. Electronic message boards, mobile apps, and other communication channels can inform motorists of speed limit changes in real-time. For the speed to be enforceable, there needs to be roadside signs that follow the MUTCD standards for regulatory speed limit signs, which includes a white background with black text and a black border.

Implementing VSL systems can be challenging due to the need for real-time monitoring, accurate traffic modeling, and effective communication with drivers. In addition, VSL systems require significant investment in technology and infrastructure. However, VSL is an effective strategy for managing traffic flow and improving safety on high-speed roads.

2.3.5 SPEED HARMONIZATION

Speed harmonization is a traffic management strategy that involves optimizing and reducing the variability of speed at which vehicles are traveling on a roadway based on current traffic conditions. The goal is to improve traffic flow and reduce congestion by optimizing the speed of traffic and minimizing disruptions.

Speed harmonization systems rely on real-time monitoring of traffic conditions to determine the optimal speed limit for the roadway. This may involve sensors embedded in the road, cameras, or other technologies. Effective communication of the target speed with drivers is critical for the success of speed harmonization systems. Electronic message boards, mobile apps, and other communication channels can be used to provide motorists with the updated target speed.

2.3.6 DIVERSION ROUTING

Diversion routing is a traffic management strategy that involves diverting traffic away from a congested roadway or area to alternate routes. The goal is to reduce congestion and improve traffic flow by distributing traffic across multiple routes. Diverting traffic to alternate routes can have safety implications, particularly if those routes are not designed to handle heavy traffic volumes.

Diversion routing systems must take into account the safety of all road users, including drivers, pedestrians, and bicyclists. By diverting traffic away from congested areas and distributing it across multiple routes, diversion routing systems can improve traffic flow, reduce travel times, and help drivers reach their destinations more efficiently.

Diversion routing systems rely on real-time monitoring of traffic conditions to determine when and where to divert traffic. Determining when to divert may involve sensors embedded in the road, cameras, or other technologies. Determining where to divert traffic could be done either dynamically or via a selection of pre-determined diversion routes from which to choose. This determination could be done by an algorithm, by an operator, or a combination via operator usage of a decision support system. These alternate routes will be required to handle the additional traffic. This may involve using parallel roadways, designated truck routes, or other options.

Electronic message boards, mobile apps, and other communication channels can be used to provide alternate routes to motorists. Information can be presented as driving directions or trail-blazer message sets at the right location to guide motorists along the diversion route.

2.3.7 SIGNAL CONTROL

Signal control can be used as an advanced strategy to address non-recurring congestion caused by incidents and congestion in conjunction with diversion routing. When an alternate route contains signals, there will likely be an increased demand in the movements through the intersection in the direction of the diversion route.

The traffic signal timing pattern can accommodate this increase in demand by providing additional time within the cycle to those movements. This can improve the level of service for the transportation network, allowing the congestion to dissipate faster in both the corridor from which traffic is being diverted and the diversion routes onto which traffic is being diverted.

2.3.8 BUS BRIDGING

Bus bridging is a transportation service that involves using buses to replace rail service on a specific portion of a transit system's route. It is typically used when there is a disruption or maintenance on the rail line that requires it to be closed or limited in service. By using buses to provide temporary service on the affected portion of the route, transit agencies can ensure that riders are able to reach their destinations while the necessary maintenance or repairs are completed on the rail line.

Bus bridging requires coordination between the transit agency and the bus operator to ensure that adequate buses are available and that they are dispatched to provide the required service. Bus bridging service should be designed to meet the needs of the affected ridership, taking into account factors such as ridership levels, the length of the disruption, and the availability of alternate routes or modes of transportation.

Bus bridging service should be designed with safety in mind. This may include additional signage and barriers to ensure that riders are safely guided to the temporary bus stops and that buses are able to safely navigate the area.

Bus bridging can be costly for transit agencies, as it requires additional buses and operators to provide the service. Transit agencies must weigh the cost of providing bus bridging service against the potential impact on ridership and the overall transit system.

2.4 INTEGRATED CORRIDOR MANAGEMENT POLICIES AND CONSTRAINTS

Traffic management operating procedures and traffic control devices have a significant impact on safety for the traveling public. This responsibility has led to policies and constraints on traffic control devices.

2.4.1 PUBLIC SAFETY

Public safety is one of the top priorities in traffic engineering and operations. An ICM strategy cannot invoke something that would be unsafe. Even implementing a response that would have worse performance than not implementing the response could lead to congestion and rear-end collision. Traffic engineering is far more complex than any single algorithm can support alone, and as such, engineering judgement and oversight is an important requirement for ICM.

2.4.2 ENGINEERING OVERSIGHT

Public safety and machine error due to lack of intuition can be mitigated by having an operator in the loop to review and approve recommended actions by the system. ICMS system recommendations are presented to the operator for review. The operator reviews and modifies the recommendation before invoking it in the traffic management system.

A roadmap for implementing ICM typically begins in phases. Initially, the ability for an operator to monitor the facility and use human judgement to determine and invoke one or more of these strategies via manual operating procedures with existing systems is the first phase. Once procedures are in place governing the ICM strategy, pieces of the operation can be evaluated for automation and improvement. Monitoring the system for incidents, optimizing the system for performance, identifying and analyzing a specific strategic response, and actuating the response are all candidates for automation. An operator typically remains part of the workflow to at least audit the suggested response and make the final decision before the response is invoked.

2.4.3 SAFE TIMING PATTERNS

Various policies are in place for traffic signals to maintain safety. There are minimum and maximum limits on yellow and red intervals for safety. There are also limits on orders of phases based on signal heads to protect safety. For example, with a 5 section signal head, a permissive left turn (green ball) must end at the same time as the opposing through phase. This prevents the left turn yellow trap where a driver turning left permissively misunderstands the ending of their permissive phase as the ending of the conflicting phase as well. Finally, there are hardware constraints that prevent unsafe indications from being displayed. The conflict monitor, in some cases called a malfunction monitoring unit, automatically prevents phases that conflict from getting a green indication at the same time.

When the operating traffic signal timing pattern is changed, it initially may not be in synchronization with the other intersections it is in coordination with. This can impact the performance initially, take several complete cycles for the pattern to get back into synch with the coordinated corridors, and then the performance benefit will be realized. Thus, once a timing pattern is changed, it should not be changed again for at least 3 to 5 complete cycles.

When a pattern changes, the current offset and new offset may differ. Phase lengths are adjusted through a transition algorithm to slowly change the current offset to the new offset. There are two main methods for transition: long- and short-way. Long-way transition increases the phase lengths to change offset. Short-way transition decreases phase lengths to change offsets. Each transition method has a maximum difference that can be set on each phase (generally defined as a percentage) and must continue to serve the minimum times defined for each phase.

A signal controller data only supports up to a certain number of traffic signal timing patterns. This can be between 32 and 256. Furthermore, there's a limit to the number of patterns that can be feasibly maintained. This impacts some methods of ICM where custom patterns are defined for particular events. Controller limitations on number of patterns available may limit the flexibility associated with these corridor management strategies.

2.4.4 DATA REQUIREMENTS

The following are some of the data requirements for achieving the objectives described above:

1. High resolution traffic detection: Traffic detection data is available in real-time to the traffic signal controllers. In some locations, this data is used for simple actuation for signal timing. In other locations, adaptive signal timing is used which has more dynamic changes to signal timing based on the real time detection data
2. Turning movement counts: In traditional, periodic, signal retiming, turning movement counts are used to evaluate new signal timings.
3. Travel times: Travel time data is generally used as an evaluation criteria for signal timing. The purpose of the roadway is important to consider when evaluating based on travel time. For roadways whose primary purpose is access, travel time along long segments of the roadway may be less critical to corridor operations than the performance of side street turning movements and shorter segments of travel time.
4. Performance measures: High resolution controller data can be used to create performance measures such as arrivals on green/red, platoon ratio, split failures, etc. These are discussed in chapter 3 section 5.
5. Traffic Incident detection and status: If traffic incidents are detected, a complex signal timing system could adjust in real time to accommodate the change in traffic demand.
6. Traffic control device status: Real time communication with traffic control devices can enable faster dispatch and repair when the devices fail.

2.4.5 CROSS-JURISDICTIONAL CHANGE MANAGEMENT

ICM can spans multiple jurisdictions, i.e. limited-access facilities and state-owned roadways may be within the jurisdiction of the state department of transportation (DOT), while the arterial networks could span multiple cities and counties, each operating facilities within their geographic region. Furthermore, operations can be outsourced and performed under contracts or memorandum of understanding or other agreements. All of these jurisdictions are needed to partner together for a successful ICM program, while all of them will have their own set of policies constraints that need to be followed.

2.5 CONCLUSIONS

In conclusion, this chapter has provided an in-depth exploration of various aspects crucial to transportation management systems. It has elucidated the fundamental role of traffic signals, emphasizing their importance in optimizing traffic flow and coordination between intersections. The discussion on time-space diagrams has underscored their utility in evaluating signal timings and assessing intersection coordination effectively.

Furthermore, the chapter has delved into the operational modes of traffic signals, emphasizing the significance of stages and phases in signal timing. It has also addressed the management of limited access facilities, highlighting key aspects such as traffic management, incident response, toll collection, and the integration of intelligent transportation systems (ITS) for enhanced efficiency and safety.

Moreover, the exploration of advanced traveler management system strategies, including Managed Lanes, has underscored their benefits in improving traffic flow, reducing congestion, and providing more efficient travel options for drivers. The chapter has also emphasized the importance of integrated corridor management policies and the paramount role of public safety in traffic engineering and operations.

In summary, this chapter advocates for meticulous coordination, engineering judgment, and oversight in traffic management to ensure safe and effective operations. It stresses the integration of operator review and approval in Intelligent Corridor Management System (ICMS) recommendations to mitigate risks associated with machine error, ultimately striving for optimized traffic flow, enhanced safety, and reduced congestion in transportation systems.

3 Integrated Corridor Management System

Abstract: An integrated corridor management system (ICMS) implements the ICM strategies using automated functions powered by data in real-time. While some functions are automated, many functions play a part of a workflow using a transportation operator to review, revise, confirm, and finally activate a strategy. The ICMS will require a robust data management subsystem to collect and manage data to power other subsystems. Other subsystems implement business logic around one or more traffic systems or advanced traveler management strategies. Several interfaces exchange information between operators and other actors in the system throughout the workflow to implement the strategies.

3.1 INTRODUCTION

Transportation management in modern urban environments requires sophisticated systems capable of integrating data from various sources to make informed decisions in real-time. An Integrated Corridor Management System (ICMS) stands at the forefront of this effort, implementing strategies that optimize traffic flow and mitigate congestion through automated functions fueled by real-time data.

The ICMS operates within a framework where automation is complemented by human intervention, ensuring that strategies are not only efficient but also adaptable to dynamic transportation scenarios. This integration of automated processes and human oversight forms the backbone of an effective ICMS.

In this chapter, we delve into the intricacies of an ICMS, exploring its architecture, key components, and operational processes. We begin by dissecting the system into three main subsystems: data management, decision support, and user interface. Each subsystem plays a crucial role in the overall functioning of the ICMS, from collecting and processing data to facilitating human interaction and decision-making. These subsystems are depicted in Figure 3.1.

The data management subsystem serves as the backbone of the ICMS, providing a robust foundation for collecting, managing, and distributing data from various transportation networks and sources. It employs a combination of data stores, pipelines, and extractors to ensure the availability and accessibility of both historical and real-time data.

Next, we examine the decision support subsystem, which leverages the wealth of data collected to generate strategic responses to transportation incidents and congestion. Through a semi-automated process, the decision support system evaluates current conditions, predicts outcomes, and presents candidate response plans for human review and activation.

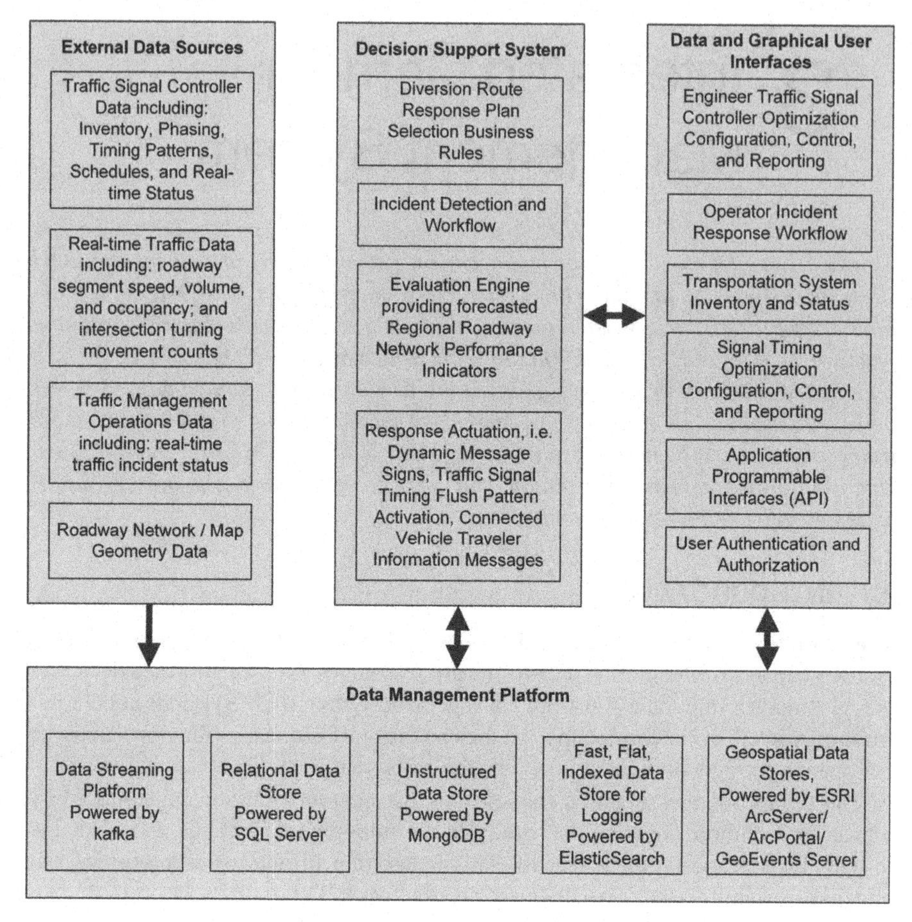

Figure 3.1: Integrated Corridor Management System Architecture Diagram

The user interface subsystem acts as the bridge between the ICMS and its operators, providing a graphical interface for interaction and decision-making. Operators can visualize transportation data, review response plans, and oversee the implementation of strategies through intuitive tools and notifications.

Furthermore, we explore operational processes within the ICMS, such as traffic incident response, signal timing optimization, and mitigation of non-recurring congestion. These processes illustrate how the ICMS dynamically adapts to changing transportation conditions, employing a combination of automated algorithms and human expertise to optimize traffic flow and enhance mobility.

An Integrated Corridor Management System represents a paradigm shift in transportation management, harnessing the power of data and automation to create more efficient, resilient, and sustainable urban mobility networks. By integrating cutting-edge technology with human ingenuity, the ICMS paves the way for smarter, safer, and more responsive transportation systems in the modern age.

3.2 SYSTEM

An ICMS relies on input data from traffic management systems to accomplish its goals and objectives. An ICMS may interface with existing data systems and control systems, providing high level functionality but relying on other systems to physically collect data and invoke changes to the transportation network. Several of these systems are described in Chapter 1, and include, the traffic signal ATMS, the freeway management system, and data feeds such as traffic conditions.

Internally, the ICMS can be organized into distinct subsystems, divided into three categories: one for collecting and managing data, one for processing data, and one for interacting with users.

3.2.1 DATA MANAGEMENT SUBSYSTEM

The data management subsystem will serve as the repository for a range of datasets utilized and generated by the ICMS and external systems. Data connectors will be created to ingest transportation network data from diverse providers and sources into the data platform. Each data source and device will provide an application programmable interface (API) specification describing the interface from which to connect and receive data. The data management subsystem will subscribe to data from each of these interfaces to ingest into one or more data pipelines, make these pipelines available to other subsystems and components within the ICMS, and may also make these data sources available to external systems via an API of its own. Historical and near real-time data must be available to an ICMS. Processes must be in place to extract data from devices and other sources, transform the data to ensure quality and consistency, and load the data in a way that is accessible to the ICMS algorithms. Some data sources are slowly changing, such as basemaps, transit routes and schedules. These data sources can be collected and loaded manually if an automated feed is not available. Other data sources are continuously changing across time and space, thus routines that continuously poll or subscribe and continuously receive data must be in place. A data catalog, or a list of data sources needed by the project is identified. Each data source is described in terms of how to access it, how to store it, and how to use it within the system. Other attributes such as the format and velocity of the data are identified as additional attributes and used to determine the best data store for that particular data source. Data stores are then deployed and configured to store the data. There are a few different types of data stores used to accommodate the different types of data available.

- A relational database management system (RDBMS) accommodates highly structured data.
- A document database accommodates semi-structured data that is in a document format rather than being normalized into relational tables. AVRO and JSON are formats typically used for streaming data and for document formatted data.
- A distributed file system is useful for storing files without any structuring or transforming of the information into one of the other data stores.

A data pipeline platform is established to facilitate distribution of data to multiple consumers with efficiency and simplicity. This is especially helpful for real-time, continuously updating data. Kafka, and other types of message queuing systems are good candidates for implementing the data pipeline. Data velocity and volume are important considerations in selecting a pipeline. If the data velocity and volume are high, the ability to scale horizontally across multiple severs is a needed capability of the data streaming platform. A data extractor is developed to extract the data from each data source. The extractor then puts the data into the data pipeline for other components to receive the information. The data is also stored in the data store for later access. Collectively, this system uses a combination of data stores, a data pipeline, data extractors, and a data catalog to manage the data sources associated with the system.

3.2.2 DECISION SUPPORT SYSTEM

The decision support system (DSS) provides a semi-automated process that evaluates the current and predicted conditions of the regional transportation network to generate strategic, candidate responses to both recurring and non-recurring congestion incidents. Operators then use judgement to make a final determination of which response to select, fine tune, and invoke. Internal components of a DSS can include a database of pre-defined response plans from which to draw candidate responses plans for a particular incident, a rules engine used to select one or more response plans based on incident data, and a predictive simulation engine. The rules engine will construct and select response plans for assessment, model the anticipated outcomes of these plans, evaluate and score them, coordinate with operators and local agency maintainers via the user interface subsystem, and implement approved response plan actions when needed. Following the activation of a response plan, the DSS will persistently monitor conditions for any differences, such as further upstream congestion queue growth or queue dissipation, and recommend adjusted responses or deactivation of the response as necessary.

Optimizing signal timing across multiple intersections requires the following concepts:

1. *Corridor Definition:* Before intersections can be coordinated and optimized, the corridors must be identified. The corridor is identified as the path of the major traffic flow, loosely defined as the largest movement volumes along adjacent intersections. Defining it this way allows for a corridor to turn onto other roadways. Corridors can vary and are thus defined in a recurring fashion for the same time periods and days of the week. A traffic engineer can define these corridors using volume and origin/destination information. Machine learning approaches to automating the corridor definition will be presented later.
2. *Timing Pattern Definition:* For each corridor defined for each time of the week, timing patterns are defined for these recurring conditions according to the principles presented in chapter 1. Additionally, special timing

plans called flush plans are also developed to accommodate a heavier travel demand through the movement that coincides with anticipated diversion routes.

3. *Response Plan Set Definition:* Diversion routes are defined to route traffic around potential congestion queues along a corridor of interest. This is done in two parts – scenario links and diversion routes. The segments for which to route traffic around is defined as a scenario link, whose starting point is an exit off of a corridor and whose ending point is an entrance back onto the corridor. Scenario links can be combined to accommodate congestion queues that traverse beyond a contiguous exit-entrance combination. Diversion routes are defined with their path through the network from an exit at the beginning of a scenario link, through the network, to an entrance back onto the corridor. Diversion routes can be assigned to one or more scenario links for routing around congestion within the scenario link. Diversion routes also include additional elements that implement the strategies in addition to the re-routing of traffic, including signal "flush" timing patterns to accommodate the increase in traffic demand from the diverted traffic, DMS messages and CV TIMs to direct traffic onto and through the diversion route, and other strategies including ramp metering, express lanes pricing, and emergency shoulder use.
4. *Incident Detection and Response Plan Selection:* Operationally, the ICMS receives traffic information to detect incidents and associated congestion to down select the scenario links and their associate response plans as candidates to be considered for selection and usage for the incident and associated congestion. There can be additional parameters involved, but generally, the incident causing a capacity reduction and congestion is first used to find a matching set of adjacent scenario links, then the diversion route response plans associated with that scenario link are considered as a set of candidate response plans. These candidate response plans are evaluated either by the mesoscopic simulator for their forecasted performance in terms of travel time and delay impacts on the region of the network. Once evaluated, the set of candidate response plans can be ranked according to their measure of effectiveness forecasted for their impacts to the transportation network. With the rankings, the system can either proceed with invoking them, or, based on the policy/constraint they can be presented to an operator or other stakeholder for review, fine tuning, and approval before invocation.

3.2.3 USER INTERFACE SUBSYSTEM

The graphical user interface (GUI) provides interaction with operators and agencies to support coordination and human operator judgement as part of the overall workflow. Users can use the GUI to view the entire dataset of transportation information and interact with the system including the current transportation conditions, device status and detection data, detected incidents, candidate response plans including and ranked by performance impacts, and status of the response itself. The operator can review, select, revise, and invoke the response plan. The GUI also provides notifications through text, emails, and mobile applications to notify stakeholder agencies of events and response plan actions that need operator attention to review, select, fine-tune, invoke, and manage a response plan. Finally, the GUI also supports post event analysis of prior incidents and response plan events and allows the traffic engineer to configure the response plan set available for drawing candidate response plans, along with other engineering tools that will benefit the transportation network such as signal optimization and other ICM strategies in addition to the incident response operation.

3.3 TRAFFIC INCIDENT RESPONSE

3.3.1 LIMITED-ACCESS FACILITIES

This operational process monitors the traffic information to detect incidents on limited-access facilities and determines an adjacent corridor flush plan as part of an incident response plan. To support the determination, the ICMS determines the measure of effectiveness of a set of candidate response plans. This determination can be done using an online, mesoscopic simulation of the affected portion of the transportation network to produce performance measures of the network. This approach allows a wholistic view of the network rather than just considering treatment of the symptoms. The process then coordinates with the ICMS operator and any other decision maker from which approval is required before invoking the response plan. The process continues to monitor the incident and environment conditions, makes adjustments to the response plan, and finally deactivates the response plan when conditions return to normal.

3.3.2 ARTERIAL FACILITIES

This operational process monitors the turning movement counts, the signal performance measures data, and quality processed traffic detection information output for an increase in queue length and/or travel time over a configurable percentage of historical value, determines a better-optimized corridor signal timing plan set from existing plans, determines the measure of effectiveness of the signal timing plan set using a mesoscopic simulation of a reduced network, coordinates with FDOT and/or local agency maintainers for approval, invokes the response plan, continues to monitor the incident and environment conditions, makes adjustments to the response plan, and finally deactivates the response plan.

3.4 PERIODIC SIGNAL TIMING OPTIMIZATION

Recurring congestion, or rush hour, is the oversaturation of vehicles beyond what the roadway network can handle. Typically, this results in delay as compared to free flow conditions. At intersections, this can manifest itself into split failures. This offline process optimizes the network by considering performance, i.e. volumes, turning movement counts, or saturation rates, for each movement of each intersection within a corridor or a signal in isolation, groups the intervals into contiguous intervals based on similarity of their performance, clusters similar groups of contiguous intervals, calculates new optimal timing plans for the cluster of contiguous groups, calculates the percent improvement and applicability over the plans currently in use, presents optimal plans and metrics and related data to the signal timing engineer for review and timing plan implementation. The engineer can view the optimal plans and related data, make adjustments and new plans, request reduced mesoscopic simulations, and provide signed and sealed plans for approval and implementation.

3.5 OPTIMIZE NETWORK FOR RECURRING CONGESTION

Additionally, a signal timing optimization engine can improve traffic flow through a corridor by optimizing the traffic signal timing patterns to be in coordination along a corridor. An ICMS can invoke a signal timing optimization engine and provide data including turning movement counts. The engine will output optimized timing patterns for review by a signal timing engineer, who can make fine tune adjustments, run the modified timing patterns through the simulation engine, and deploy the signal timing patterns into the traffic management system to be used in operations.

3.6 MITIGATE IMPACTS FROM NON-RECURRING CONGESTION

Non-recurring congestion typically refers to an emergent reduction of capacity oftentimes caused by a vehicle crash or disabled vehicle. There are other situations that cause non-recurring congestion including a special event such as a concert or sporting event, and even congestion that exceeds the historical norm. The ICMS will detect non-recurring congestion, and respond to it by diverting transportation demand to other modes or facilities with available capacity. This response will be presented as a decision support for operator review, fine-tuning, and invocation. Pursuant to the system goals, the ICMS must accomplish several objectives, including[1]:

1. Receive real-time and historical data from traffic and transportation-associated systems and operations.
2. Examine infrastructure status data to assess the accessibility of infrastructure components and/or systems for incorporation into corridor enhancement strategies and regional response plans.
3. Provide users real-time updates on the status of devices and the performance of both roadway and transit networks.
4. Examine gathered data to assess transportation performance, identify issues, congestion, and incidents, and explore potential corridor improvement strategies and responses to traffic events and congestion. Strategies and responses may encompass, but are not confined to, the following:
   - Responder status, dispatch, coordination, activities, and related timestamps for arrival and clearance activities.

[1]https://ops.fhwa.dot.gov/fastact/atcmtd/2017/applications/pinellascounty/project.htm?

- Selection of a diversion route or set of diversion routes having available capacity around a traffic incident.
- Selection of a timing pattern to handle the increased demand along diversion routes.
- Selection of specific ramp metering rate to slow the rate of vehicles entering the limited access facility upstream of a traffic incident and to allow an increased rate of vehicles to enter the facility downstream of an incident.
- Opening of the shoulder for use as a travel lane.
- Reducing or eliminating express lanes to encourage more vehicles to choose the express lanes.
- Transit rerouting and bus bridging.
- Diversion route messaging to motorists to inform them of the recommended diversion route to take. This can be in a combination of the following channels:
    - Dynamic message signs (DMS)
    - Highway advisory radio (HAR)
    - Advanced Traffic Information System (ATIS)
    - 511 phone system
    - Website
    - Mobile application
    - In vehicle navigation systems
    - Connected vehicle traveler information message (TIM)

5. Assess the potential advantages of implementing corridor improvement strategies and their associated response plans through real-time and offline simulation.
6. Evaluate the effects of implemented corridor improvement strategies and their associated response plans in real-time and offline.
7. Deliver transportation-related status updates, analyses, and recommendations for corridor improvement and response plans to stakeholders in an interactive, real-time format.

3.7 CONCLUSIONS

In conclusion, the Integrated Corridor Management System (ICMS) emerges as a vital tool in the quest for efficient and sustainable urban transportation. Through its integration of automated functions with human oversight, the ICMS represents a holistic approach to managing traffic flow and mitigating congestion in dynamic urban environments.

The key components of the ICMS—including the data management subsystem, decision support subsystem, and user interface subsystem—work in tandem to collect, process, and act upon transportation data in real-time. This synergy enables the ICMS to generate strategic responses to transportation incidents and congestion, optimizing traffic flow and enhancing mobility for commuters.

Operational processes within the ICMS, such as traffic incident response, signal timing optimization, and mitigation of non-recurring congestion, further illustrate its adaptability and effectiveness in addressing the evolving challenges of urban transportation. By leveraging advanced technology and human expertise, the ICMS not only improves the efficiency of transportation networks but also enhances safety and resilience in the face of unforeseen events.

In summary, the Integrated Corridor Management System represents a transformative approach to transportation management, one that embraces innovation and collaboration to create smarter, more responsive urban mobility solutions. As cities continue to grow and evolve, the ICMS stands ready to play a central role in shaping the future of transportation, driving towards a more sustainable and equitable urban landscape.

4 Traffic Data Modalities

Abstract: Traffic datasets are at the heart of any advanced transportation management system. Datasets provide situational awareness of current conditions, derivations, and predictions necessary for systems and operators to make decisions. Data is collected and derived by several modalities for several usages, as detailed in the following sections.

The effective management of traffic relies heavily on the availability and utilization of diverse data modalities. This chapter delves into the plethora of traffic data modalities essential for understanding and optimizing transportation systems. From traditional fixed point detection methods to advanced vehicle trajectory data and incident management systems, this chapter navigates through the intricacies of various data sources crucial for traffic analysis and management.

The chapter commences with an exploration of fixed point detection mechanisms, encompassing the utilization of loop detectors, microwave detectors, and advanced video and LiDAR technologies. These detection methods serve as fundamental pillars in capturing real-time traffic information, providing insights into vehicle movements and flow patterns at specific locations.

Subsequently, the discussion extends to vehicle probe detection systems, which offer dynamic and versatile data collection capabilities by leveraging vehicle-based sensors and communication technologies. Moreover, the chapter delves into the analysis of vehicle trajectory data, elucidating its significance in understanding individual vehicle movements and behavior within transportation networks.

Furthermore, incident data emerges as a critical component, encompassing incident detection mechanisms and operational response protocols aimed at mitigating disruptions and enhancing traffic safety. The chapter also examines supplementary datasets, including third-party traffic conditions data and multimode data sources, which augment the breadth and depth of traffic analysis.

An in-depth exploration of automated traffic signal performance measures and corridor performance data follows, shedding light on methodologies for evaluating signal efficiency and corridor-wide traffic performance metrics. Additionally, the chapter elucidates the temporal horizons of traffic data utilization, ranging from long-range planning to active operational decision-making.

Finally, the chapter addresses the crucial aspect of data quality, emphasizing factors such as accuracy, availability, validation, and machine accessibility/readability. By ensuring the integrity and reliability of traffic data, transportation agencies can effectively leverage data-driven insights to optimize traffic management strategies and enhance overall system efficiency.

4.1 FIXED POINT DETECTION

Fixed point detection is a modality of collecting traffic data that involves the use of fixed sensors or devices placed at specific locations to monitor traffic parameters. They capture individual vehicle presence and movements at a location. Within the processing component of the sensor system, vehicle presence and movements are aggregated into metrics for a time interval. Typically, the time intervals are on the order of 30 seconds or a minute. This aggregated data can be referred to as bin data, denoting the time intervals as bins as follows:

1. Speed: Vehicle speed measures the rate at which the vehicle is moving, expressed in miles per hour or kilometers per hour.
2. Volume: Volume is the quantity of vehicles that pass a given point at which volume is being measured.
3. Occupancy: Occupancy is the linear percentage of the roadway covered by vehicles. In a non-interrupted flow i.e. away from intersections, occupancy can be approximated as the percentage of time during which a vehicle is present on the detector.
4. Vehicle Classification: Vehicle classification is determined when vehicle attributes are detected which include the vehicle's length, number of axles, weight, or some other configuration. Vehicles are aggregated into vehicle class bins and the bins are observed over fixed time intervals.

Among the fixed point detectors are inductive loop detectors, piezoelectric detectors, microwave vehicle radar detectors, video cameras and LiDARs installed on the roadway at traffic intersections or otherwise. These sensors are described below.

4.1.1 LOOP DETECTORS

Inductive loop detectors are sensors embedded in the road surface and detect the presence of a vehicle periodically, at a decisecond granularity. At such a fine granularity, vehicles can be distinguished from one another to allow volume to be calculated. If a second detector is placed immediately downstream, the time delay between detecting a vehicle from the upstream detector to the downstream detector can be used to calculate the speed of the vehicle. Inductive loops have an advantage that they are extremely accurate and not affected by weather. Although not directly useful for individual vehicle detection, they can be configured in several different ways to accomplish the collection of speed, volume, or occupancy. For example, at an intersection, there is a need to detect the presence of a vehicle in order to place a call into the controller.

Adjacent approach lanes do not necessarily need to be distinguished from one another to support the operational need of determining the presence of a vehicle in either lane at the stop bar. Thus, some detection configurations detect across multiple lanes, or detect a long zone that cannot distinguish multiple vehicles closely spaced behind one another.

Similarly, the detector may place a hold on the detection output for a predetermined amount of time that would be unable to detect more than the initial vehicle throughout the direction of the hold. Modern guidance for intersection loop detection providing the most flexible use of the detection would be for small zones within individual lanes to be configured, and for the controller to aggregate the detection as needed to derive calls needed for operations.

A key disadvantage of these detectors is that they must be installed in the pavement which requires lane closures.

Piezoelectric sensors are similar in concept to inductive loops, except that they measure pressure changes caused by vehicle tires rather than by the electromagnetic field.

4.1.2 MICROWAVE DETECTORS

Microwave radar vehicle detection is a technology used to detect and track vehicles using microwave radar signals. It utilizes radar, which stands for radio detection and ranging, to measure the presence, position, speed, and other characteristics of vehicles on the road. By analyzing the frequency shift of the reflected signals, the radar system can determine the speed and direction of the vehicle. This is based on the Doppler effect, which causes a change in frequency when there is relative motion between the transmitter, receiver, and target (vehicle). The radar system processes the received signals to extract relevant information, such as the presence, position, speed, and sometimes even the size or classification of the detected vehicle.

Microwave radar vehicle detection offers a few advantages over other detectors, including the ability to be installed on and maintained from the side of the road without requiring road closures. However, when installed on the side of the road, lanes can be obstructed by tall vehicles in adjacent lanes.

In-pavement induction loops and zones from microwave radars at intersections are integrated into the traffic signal controller in order to support the semi-actuated operation of the intersection. The information from the detectors and detection zones are recorded by the controller at a 10 Hz (1 decisecond) granularity. The information recorded is whether or not a vehicle is present over the loop or zone. This time-series information can be processed to derive the volume and further processed with sufficient context information to derive turning movement counts and other performance measures covered in Chapter 8.

For correctly interpreting the data gathered at traffic intersections and computing performance measures at intersections, certain secondary information is needed. The detector-to-phase mapping information is the most critical requirement for interpreting the high-resolution signal and detector event data logged in a controller. Since any detector event must be understood as a vehicle arriving at some right-of-way and distance at the intersection and any phase event must be understood as an interval of time elapsed for some movement, the detector mapping process is the critical step in guiding how signal output states and vehicle arrivals are to be interpreted relative to each other. Without the provision of such information, it would be impossible to generate most of the signal performance measures from the detector and phase data independently. Other things that might be needed are the timing sheet and location of the signal. The timing sheet includes manually programmed timing plans which is unique to each signal. This is specified by the cycle plan column of the split reports generated by most ATMS systems.

4.1.3 VIDEO AND LIDAR DETECTORS

Fixed point detection can be achieved using video to emulate a detection zone similar to other in-pavement detection types. The analysis component provides a technician the ability to configure detection zones within the field of view of the video and produces volume, speeds, occupancy, and potentially other metrics as output.

LiDAR detection is similar to video detection. However instead of capturing a two-dimensional field of values to analyze, LiDAR detection generates three-dimensional spatial mapping. These values represent distance from the sensor detected by a laser scanning the field of view, similar to radar. LiDAR is more expensive than radar and is more affected by fog, rain, and dust.

4.2 VEHICLE PROBE DETECTION

Vehicle probe detection systems are advanced traffic monitoring and data collection technologies that use probes or sensors to gather information about vehicle movement and traffic conditions. These systems are designed to provide real-time or near real-time data on travel times, speed, traffic flow, and other relevant parameters. Local vehicle identification systems utilize identifiers of vehicles as they move through the transportation network. Identifiers can include:

- Bluetooth MAC address or WiFi MAC address from a smartphone in the vehicle,
- Tire pressure monitoring sensor IDs,
- Tolling tag RFID, or
- License plate or USDOT number recognized by an optical character recognition.

By monitoring the time it takes for these identifiers to pass between two detection points, travel times and average speeds can be calculated.

Location reporting systems rely on vehicles communicating their location periodically and then aggregate the location information of the vehicle to calculate the vehicle's speed and travel time between two or more locations. Global Positioning System (GPS) technology and cellular phone location services are used by a communications device in the vehicle to detect locations and transmit the location back to a central aggregation system over cellular communications network or over digital short-range radio communications. By analyzing the location data, these systems can provide information about vehicle speeds, travel paths, and travel times.

Vehicle probe detection systems can be processed to provide similar data to fixed-point detection, plus additional parameters such as actual travel times over a distance and some flow and behavioral parameters.

1. *Travel Time Data:* The time it takes for vehicles to traverse a path observed by the probe system. Vehicle probe systems can generate accurate travel time data with a vehicle participation rate as low as 5%.
2. *Speed Data:* In identification-based probe systems, speed is calculated as the inverse of the travel time.
3. *Relative Volume and Occupancy:* Probe systems can count the number of vehicles reporting at a location; however, this volume is only relative to the actual volume based on the percentage of vehicles participating as probes. A volume count could be extrapolated from historical ground truth data by calculating a coefficient from the ground truth volume or occupancy and the relative volume or occupancy data and using that coefficient to scale future relative volume or occupancy values.
4. *Origin-Destination:* Origin-destination data provides statistics on the travel patterns between a set of origins and destinations. This can be captured as a matrix containing the origins on one axis and the destination on the other axis. The values of the matrix are the counts or relative counts of trips between the corresponding origin and destination.

However, vehicle probe data can be limited by the percentage of the vehicles participating in the probe system, particularly for volume and occupancy measurements.

4.3 VEHICLE TRAJECTORY DATA

Vehicle trajectory data refers to detailed information about the movement and behavior of individual vehicles within a transportation network. It provides a comprehensive record of the positions, speeds, accelerations, and other attributes of vehicles over time. This data is collected using advanced technologies such as GPS, video cameras, radar, LiDAR, or higher-resolution vehicle probe detection systems. Vehicle trajectory data provides rich insights into traffic patterns, driver behavior, and roadway performance. It is used for a wide range of applications, including traffic simulation and modeling, congestion analysis, safety studies, travel time estimation, route planning, and the development of intelligent transportation systems. Here are some key characteristics of vehicle trajectory data:

1. *Position and Coordinates:* Vehicle trajectory data includes the geographical coordinates (latitude and longitude) that represent the location of each vehicle at different points in time. This allows for precise tracking of vehicle movements along specific routes or within a network.
2. *Time Stamps:* Each data point in the trajectory dataset is associated with a time stamp, indicating when the position measurement was recorded. This time information enables the reconstruction of the vehicle's movement over a specified time period.
3. *Speed and Acceleration:* Vehicle trajectory data captures the speed and acceleration of vehicles at each timestep. This information helps analyze driving behaviors, identify speed patterns, and evaluate acceleration/deceleration characteristics.

4. *Heading and Orientation:* Trajectory data can provide the heading or direction of vehicles, indicating the orientation of their movement. This data is crucial for understanding vehicle movements at intersections, roundabouts, or complex road geometries.
5. *Lane Assignment:* Trajectory data includes or can be used to determine the specific lane in which each vehicle is traveling, enabling lane-specific analysis and assessment of lane-changing behavior and lane utilization.
6. *Vehicle Attributes:* Additional attributes such as vehicle type (car, truck, motorcycle), size, weight, or other specific characteristics may be included in the trajectory data. These attributes allow for a deeper analysis of vehicle-specific behavior and performance.
7. *Saturation Flow Rate:* Saturation flow rate refers to the maximum number of vehicles that can pass through a given traffic lane or intersection approach during a specific time period under ideal conditions. It is typically measured in vehicles per hour per lane (vph/l) or passenger car units per hour per lane (pcuph/l). The saturation flow rate depends on various factors, including the geometry of the intersection, signal timing, lane configuration, and driver behavior. It is influenced by factors such as the number of lanes, lane width, presence of turning lanes, presence of pedestrians, signal phasing, and the level of congestion. Traffic engineers and transportation planners use saturation flow rate as a key parameter in analyzing and designing intersections and road networks. It helps in determining the capacity of an intersection and optimizing signal timings to improve traffic flow.
   The specific value of saturation flow rate varies depending on the type of intersection, signal control, and the specific conditions. Typical values for saturation flow rate range from around 1,800 to 2,400 passenger car units per hour per lane (pcuph/l) for signalized intersections in urban areas. However, it is important to note that actual saturation flow rates can be lower due to factors such as heavy pedestrian activity, high-turning movements, or other localized conditions. Saturation flow rate can be approximated using the Highway Capacity Manual, or it can be observed as the maximum vehicle count over a period of time during the active phase for a movement.
8. *Turning Movement Counts:* Turning movement counts refer to number of vehicles making specific types of turns at an intersection or a specific location. The movements of vehicles are classified as they move through the intersection as right turns, left turns, through movements, and U-turns. Turning movement count studies are conducted to gather data on traffic patterns and to understand the flow of vehicles at an intersection. This information is used by transportation engineers and planners to evaluate the performance of intersections, assess the need for traffic signal changes, analyze the impact of new developments or road modifications, and make informed decisions about traffic management and infrastructure improvements.
   Turning movement counts are used to optimize traffic signal timing patterns by normalizing the counts with the movement's saturation flow rate as a ratio and distributing the movement's active phase time proportionally withing

the cycle. When an approach lane has a single movement restriction, the volume through that lane at the stop bar of the intersection can be considered to represent the turning movement count.

Approach lanes with multiple allowed, or even possible movements will have a portion of the volume move through the intersection in different movements. Thus, for accurate turning movement counts for lanes with multiple possible movements, vehicle trajectory is needed. This can be accomplished with human observation or statistical inference from past ground truth turning movement count data, video or LiDAR-based vehicle trajectory analysis, or by placing another point detection zone at each of the egress lanes.

4.4 INCIDENT DATA

Operators at traffic management centers monitor and respond to traffic events. This operation includes a workflow by which operators enter in details of and changes to events as they occur in real-time. Each time an entry is made, it is time-stamped and can be used for performance analysis of the traffic management and response operation. An important goal of agencies is to clear incidents from the roadway as quickly as possible. The time-stamped information recorded can also be used for further analysis as traffic management automation is explored.

Traffic incidents are events occur on the roadway network which impacts traffic operations. Common traffic incidents include crashes, vehicles stopped on the roadway or the shoulder, police activity, construction, and debris in the roadway. Data available for incidents varies based on the incident type and facility type. Crashes are generally the most documented incidents with law enforcement officials filling out a detailed report on conditions during the crash and the probable cause of the crash.

On freeways with service patrol trucks, there can often be detailed data available on other incidents which occur due to the combination of reports from the service patrol trucks and from the cameras overlooking these areas which are monitored by the traffic management center. On other facilities, incident data may be more sparse, or may be missing completely, due to the lack of reports of these other types of incidents made to a central source on non-monitored facilities.

4.4.1 INCIDENT DETECTION

Incident detection can be used in real-time to speed the response to incidents. With historical data, incident detection can be used to determine when incidents may have occurred on non-monitored roadways as part of a research or study into incidents on these facilities. Methods of incident detection vary based on type of facility and the data available.

- *Incident Reports*: The common state of the practice is for the traffic management center to be notified of incidents through reports from the public, law enforcement, service patrol, and via manual review of cameras.

- *Volume Difference Incident Detection*: The simplest automated method of incident detection is based on the comparison of volumes both historically and along a roadway facility. A significant drop in volume either from historical data and/or from an upstream detector indicates a capacity-impacting incident may have occurred. This type of incident detection works best on arterial facilities, where most incidents of note are likely to have an impact on capacity. This is unlikely to be as useful on freeway facilities where some incidents may not drop capacity enough to trigger a detection.
- *Detection of Incidents via Speed:* On facilities with significant deployment of sensors such as microwave sensors, the real-time report of vehicle speeds can be used to detect an incident. A sudden drop of speed at a sensor may indicate an incident has occurred nearby. Paired with volumes, weather, and geometric data, this approach could potentially even be used to predict the occurrences of an incident up to 5 minutes before one occurs. This method generally works best on freeway facilities.
- *Detection of Incidents via Cameras:* Cameras have been broadly deployed across the roadway network, primarily for operators to observe from the traffic management center once an incident has been reported or detected. However, with advanced computer vision technologies, these cameras could be reviewed automatically and issue an alert when an incident has occurred. This method could be used anywhere with sufficient camera visibility.

4.4.2 OPERATIONAL RESPONSE

Upon the detection of an incident, a series of steps are taken to respond to the incident.

1. *Detection:* The detection time is commonly referred to as the notification time, reflecting the state of the practice to rely on notifications of incidents to begin responding to them.
2. *Verification:* For incidents reported within a camera-covered area, the next step of the process is the verification stage. During this stage, dispatchers in the traffic management center utilize the cameras to verify that an active incident is occurring, the severity of the incident, and the exact location of the incident.
3. *Responder Dispatch:* Following this, the incident is dispatched to the best resources to handle the incident. This may be a service patrol vehicle, law enforcement, fire rescue, or some combination of these responders.
4. *Responder Arrival:* In non-camera covered areas, the responder patrolling the area may be the first person to observe the event. In this case, the on-scene arrival occurs at the same time as the verification.
5. *Responder Activities:* Once responders arrive, their arrival time is recorded, and they begin performing their duties on the roadside. Service patrol vehicles perform traffic control and assist motorists with disabled vehicles, in some cases moving them off the roadway to the shoulder or a breakdown area. Law enforcement acts as additional traffic control while also taking

reports and other law enforcement duties. The fire department responds to vehicle crashes requiring extrication of people, or fire response and rescue vehicles, also known as ambulances, respond to vehicle crashes with injuries requiring treatment or transport to a hospital. Other responders may be called once the first responders are on-scene; this includes towing service providers and maintenance personnel.

6. *Lane Clearance:* Other important times in the incident response include the time when all travel lanes are clear. The incident may still be ongoing, but all typical travel lanes are open. This may be immediate for incidents starting on the shoulder but may take some time for incidents that began within the roadway.
7. *Scene Clearance:* Responders and disabled vehicles at the scene still have an impact on traffic when they are moved to the shoulder or even if they are off the shoulder but still at the scene. Other motorists slow down and tend to observe the scene instead of driving at full speed. The clearance of the entire scene is thus the final step in the incident timeline to facilitate the roadway returning to normal conditions.

4.5 OTHER DATASETS

4.5.1 THIRD-PARTY TRAFFIC CONDITIONS DATA

Third-party traffic conditions dataset providers collect location information of vehicles from vehicle fleet's cell phones, in-vehicle dispatch and navigation units, and other data sources; process the data; and transform the data into a data model that can be shared as a data feed through an API consumed by agency traffic management systems. This information can be made available in real-time for operations, as well as collected or provided as a historical dataset for analysis. The types of information provided by third parties typically range from travel times and speed, to more sophisticated analytics including profiles of volume (or relative volume) and performance indicators of roadways and intersections.

4.5.2 MULTIMODE DATA

Multimodality is an important emphasis of ICM, as an important way to shift demand away from the primary modality of passenger vehicles, particularly of single occupancy. Data from transit operations is important in determining the performance of the transit system, and the available capacity in mode-shift operations. A couple of key transit datasets include the following:

- Passenger Ridership: Ridership is the number of passengers in the transit vehicle. This information can be obtained via a sensor mounted in the vehicle and transmitted back to the central system in near real-time. This information can be used to plan routes and schedules, to detect when vehicles are full and riders have to wait for the next vehicle, and to determine available capacity for mode-shifting operations.

- Automated Vehicle Location: This provides the real-time location of vehicles. This information can be obtained by a vehicle-mounted communications device capturing the location from GPS or other location services and providing it back to the central system in real-time. This information can be used in a variety of ways, including determining the performance of the vehicular movements, delays, travel times, travel time reliability, and wait times ahead of stops.

4.6 AUTOMATED TRAFIC SIGNAL PERFORMANCE MEASURES

Automated Traffic Signal Performance Measures (ATSPM) uses high-resolution signal controller data logging to support performance-based maintenance and operations strategies that improve safety and efficiency while cutting congestion and cost. ATSPM performs continuous performance monitoring supporting operational improvements including signal retiming based directly on measured performance without dependence on software modeling or manually collected data.

A high-resolution traffic signal controller, when enabled, logs every detection input at 10 Hz along with controller state information such as phase transitions into a .dat file. The ATSPM software parses the raw file into a series of controller event records with a timestamp, signal event code, signal event parameter, signal IP and a location identifier or controller ID. There is static data about the intersection that is manually entered into the ATSPM central user interface that is referenced and used to support analysis performed on the controller event logs. For example, a controller ID is linked to intersection name and a detector channel is linked to an intersection approach direction, movement, and distance from the stop bar. While this configuration information can be entered manually by a traffic engineer with knowledge of the intersection design, this information can be inferred from the data. With this configuration information, the ATSPM system can then perform calculations and aggregations of the data, and then produce performance measures information from the data that is useful for identifying specific performance issues or opportunities to make improvements to the system including device maintenance, signal timing patterns, and coordination.

The high-resolution controller event logs are produced by the data logger on each signal controller. Raw log files are extracted by ATSPM log extractor via the file transfer protocol (FTP) server hosted by the signal controller. The ATSPM log extractor then uses vendor specific decoders to parse the log files readable data into a comma separated values (CSV) file. The ATSPM system's data ingester module then inserts the processed CSV data into a common table in the ATSPM database. The ATSPM application uses controller event log data from ATSPM database to calculate various performance metrics and produce visual charts for traffic operation staff and public in response to user requests from the ATSPM web application. Visual charts are then used by traffic operation staff to identify specific performance issues or opportunities to make improvements to the system including device maintenance, signal timing patterns, and coordination. In addition to the charts produced by the ATSPM web applications, the ATSPM system provides event log data to other systems

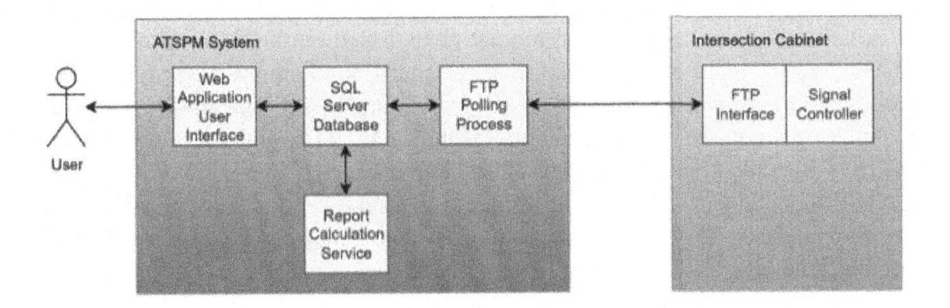

Figure 4.1: ATSPM Data Flow

that convert the log data into real-time, continuous, volumes or turning movement counts for use by the SunGuide software and the R-ICMS software. These data flows are depicted in Figure 4.1.

The primary input to ATSPM are the controller event logs. These logs convey information by specifying an event code along with a parameter. Each of these records is labelled with the controller ID and a timestamp to identify the controller and time at which the record was produced. There are a few primary event codes of interest that provide the following information:

- *Detection Call Status:* 10 Hz sampling of whether a vehicle was present over a detector zone or not. A sequence of this detection information can be analyzed to determine vehicle counts, speed between two detectors, and many signal performance measures discussed in the next section.
- *Phase Termination Status and Behavior:* The actuated signal controller implements a timing plan in response to actuated detection to make adjustments impacting the phase transition behavior. This information on the phase transition and general controller operations status/status changes includes the signal event codes. This information also supports the calculation of signal performance measures.
- *Faults and Error Codes:* Anticipated types of errors can be logged to help alert and diagnose issues with the controller that can be resolved through maintenance. Stuck pedestrian call buttons or side street detectors that are stuck in a call mode can be problematic to the intersection operation and are required to be maintained by the Traffic Signal Maintenance and Compensation Agreement(TSMCA).

The high-resolution data logging requirements shown in the FDOT traffic controller supplemental requirements are based on the November 2012 Indiana Traffic Signal Hi Resolution Data Logger Enumerations. The supplemental requirements list the six vendors [Econolite, Siemens, PEEK, Cubic (Trafficware), McCain, Q-Free (Intelight)] that capture the mandatory and optional enumerations. A complete list of the current FDOT high resolution data logging requirements is shown in Supplemental

Requirements, SR-671-2-01 and SR-671-2-02-2070 [1]. FDOT is currently exploring the expansion of the supplemental requirements based on the recent update of the Indiana Traffic Signal Hi Resolution Data Logger Enumerations August 2020. The change could introduce 45 optional and 24 mandatory events. The proposed enumerations are shown in the referenced High Resolution Data Logging Proposed Changes for FDOT Specification from 2020 vs. 2012 Indiana event codes.

ATSPM contains a variety of performance measures reports and charts for traffic operation analysis. These reports and charts, and how they can be used, are summarized in Table 4.1.

ATSPM will not be able to produce some charts if the intersection is not instrumented with the detection required for that chart's calculations. There are instances where intersections were designed and instrumented with only the detection necessary for the intended operation. A typical example of this is a semi-actuated intersection having only side-street and left-turn presence detection since the main street is always green in the absence of any other calls from the side street or left turn lanes. The signal performance measures charts able to be produced for this type of instrumentation will be limited unless additional detection is installed to cover the main street through lanes. Other detection design patterns that are limiting include having detection that spans multiple lanes, or that spans a long distance, detector loops configured as a presence and hold behavior, or not having advanced detection, all of which prevent the vehicle detection resolution needed for counts and other measures. Furthermore, there are legacy controllers that do not make signal events logs available at all to the ATSPM system.

4.7 CORRIDOR PERFORMANCE DATA

There are a variety of corridor performance data types which can be calculated through various sets of transportation data. Each type of performance data can inform a different attribute of the roadway.

1. *Travel Time Reliability:* Travel time reliability is a measure of the variability of a corridor's travel time. No two days are exactly the same. Some days have higher volume, some have lower volume, some have incidents, some have weather, some have work zones, and more that impact traffic conditions. In general, corridors which are closer to capacity on a typical type of analysis day have lower reliability as each of these traffic impacting conditions can push capacity below demand and cause traffic breakdown conditions. However, some facilities also experience poor reliability due to high incident rates and other conditions.
   Travel time reliability is often measured using a travel time index. This is the travel time divided by a base travel time. The base travel time may be based on an HCM based analysis or may be determined from the data as the free-flow travel time. The travel time in the numerator can be a variety of

[1] https://www.fdot.gov/traffic/traf-sys/product-specifications.shtm

Table 4.1: ATSPM Reports and Charts.

Report Name	Description	Usages
Purdue Phase Termination	Reports how phases terminated each cycle over a specified time period	Detector health monitoring, i.e., side street detectors that are stuck in a constant call can cause irregularities in the pattern of phase terminations and could be identified
Split Monitor	Combines split duration (in seconds) and programmed splits with the phase termination	Assess traffic signal operations
Pedestrian Delay	Shows how long a pedestrian must wait to receive a walk indication after pushing the button; summarized by phase for every cycle during a specified time period	Verify pedestrian service is operational Determine pedestrian demand Evaluate responsiveness and level of service to pedestrians
Preemption Details	Shows preemption behavior and including the input calls, max out, and the dwell time, track clear, time to service, including delay associated with preemption calls during the specified time period	Verify preemption service is operational Determine preemption demand Evaluate impacts of preemptions on other performance measures when analyzed with other reports
Purdue Coordination Diagram	Depicts vehicle arrivals compared to the signal state (i.e., green, yellow, red) for each cycle of the coordinated phases. Requires advanced detection 330 to 400 feet from the stop bar	Evaluation of progression quality along a coordinated corridor Evaluate detailed coordination related performance measures within an intersection
Purdue Link Pivot	Evaluates multiple signals along a coordinated corridor and optimizes offsets using predicted arrivals on green. Requires advanced vehicle detection	Evaluate coordination-related performance measures throughout a corridor Obtain recommendations for adjustments to offsets
Turning Movement Count	Reports vehicle volumes by lane, and the sum of volumes by lane group. Requires lane-by-lane detection	Analyze traffic demand and determine saturation rates Optimize signal timing pattern splits Determine available capacity for diversion routing
Purdue Split Failure	Reports split failures, which indicate how often vehicles are left unserved at the end of a phase. Displays Green Occupancy Ratios (GORs) and Red Occupancy Ratios (RORs) Split failures occur when both GOR and ROR are both greater than 80%. Requires presence detection at the stop bar	Analyze performance related to level of service of the intersection Indicates need for signal re-timing analysis of coordination, splits, or capacity adjustments
Approach Volume	Displays peak hour volumes for each direction of travel. Requires count detection at the stop bar or in advance of the intersection	Analyze corridor demand and saturation during peak hour Optimize signal timing patterns
Approach Delay	Estimates total delay and average delay per vehicle for each phase	Evaluation of progression quality along a coordinated corridor
Arrivals on Red	Depicts vehicle arrivals when the signal state is red for each cycle of the coordinated phases. Requires advanced detection 330 to 400 feet from the stop bar	Evaluation of progression quality along a coordinated corridor. Evaluate detailed coordination-related performance measures within an intersection
Approach Speed	Reports speed limit, average speeds, and 85th-percentile speeds. Requires speed detection in advance of the stop bar	Evaluation of when speeds change significantly and when speeds can be used as a proxy for travel time

different percentiles depending on what the user is interested in. Common percentiles are 85th, 90th, and 95th percentile. Buffer index is another factor used to report reliability. Buffer index is the additional time required to be added to a trip to ensure reliability. It is generally the travel time index minus one expressed as a percentage. For example, a travel time index of 1.4 would correspond to a 40% buffer index.

2. *Delay:* Delay is another common corridor performance metric. Delay can be reported for a corridor overall, or for individual movements at an intersection. For corridor delay, generally travel time data is used to compare the free flow travel time to current travel time. The difference between the two is the delay measure. Delay for individual movements is more complicated to determine directly using field data. Often, the HCM methodology is used to supplement this performance measure. Using the demand values from field data and the signal timing parameters, the HCM method can be used to calculate delay per movement.

4.8 TRAFFIC DATA HORIZONS

The time horizons can be grouped into three categories: long-range planning, medium-range operations, and real-time operations. These different purposes have different requirements for the data and sometimes these use different sources of data as well.

4.8.1 LONG RANGE PLANNING

Long range planning includes the activities which Metropolitan Planning Organizations (MPO), Transportation Planning Organizations (TPO), and state DOT planning and preliminary engineering offices are responsible for. These activities include regional demand modeling, long-range transportation plans, work program development, project identification, and preliminary engineering and environmental analysis work.

The data needs for this work include wide-scale daily demand information, peak hour demand percentages, and directional distribution of traffic. These are often abbreviated as AADT, K factor, and D factor. Where AADT stands for the annual average daily traffic, or the total traffic volume per day averaged over every day of the year. The K factor is the forecast year peak-to-daily ratio, or the percentage of the daily traffic which occurs during the peak hour. The D factor is the directional factor which describes how much traffic is going in each direction on a roadway. The D factor can be a daily D factor or may be a peak hour D factor.

In general, this data is processed and updated on an annual basis by the state DOT. This process involves ingesting data from count stations across the state, both permanent and temporary count stations. Permanent count stations are installed in-ground or above-ground and report traffic for the entire year. Temporary count stations are placed at particular locations for one to three days generally. The data from the temporary stations must be adjusted by a seasonal factor to calculate an AADT value. Seasonal factors are often determined using nearby permanent count station data.

Other long-range planning data includes probe data sources which have been summarized into overview statistics. For example, in Florida, FDOT generates the Sourcebook yearly. This includes statistics such as average travel speed, vehicle hours of delay, and other factors at the state and district level, as well as at a segment-level basis.

Long-range planning data is the most abstract of the data horizons due to the processing involved in data analysis. This abstracted data is easily utilized in planning-level approaches due to the preprocessing done by the state DOT but doesn't contain the details needed for a detailed analysis. The update horizon of this data (generally yearly) also doesn't allow for medium-range or real-time operations.

In general, the time scale of the data in this horizon is at the level of a single average day, and the spatial scale of the data in this horizon is at a segment level. On freeways, these segments generally run between each ramp. On arterial facilities, segments can often be long enough to cover multiple signalized intersections.

4.8.2 MEDIUM SCALE PLANNING

Beyond the abstracted long-range planning data available, in some cases more detailed medium-range operational data is needed. Often, this horizon includes the unprocessed data from the long-range planning horizon. There is often minimal processing conducted which removes poor data from the dataset. This allows for more detailed information to be determined from the data. Perhaps volume profiles over the course of the day can be determined (rather than simply a single K factor) for use in a simulation project.

Or probe data sources can be used to generate a time-of-day speed contour on a roadway to determine operational conditions throughout the day.

There are many sources of data that can be used for medium-range operations, but generally the data is less processed compared to the long-range planning data. In many cases, the data for medium-range planning may need to be collected specifically for this purpose. In general, the time scale of the data in this horizon is in the 15-minute to hourly range, and the spatial scale of the data in this horizon is at an individual intersection, and segment level.

4.8.3 ACTIVE OPERATIONS

Real-time operations refer to the active ongoing TSM&O strategies and their data needs. In this data horizon, latency is the key factor in data strategies. Due to this, minimal-to-no processing can be conducted on the data before it can be analyzed.

Systems in this data horizon need to be fault-tolerant and capable of handling bad or missing data. In particular, these systems should be able to ignore incoming bad or missing data on a real-time basis. This differs from the medium-scale operations work where full days of data can be reviewed and processed to determine missing data.

Data for real-time operations can come from systems such as ATSPM, probe data, microwave sensors, and other permanent field equipment. Data can also come from operators who can enter information about incidents and crashes which are ongoing. In general, the time scale of the data in this horizon is in the range of every 10th of a second to every 3 minutes, and the spatial scale of the data in this horizon is at the individual intersection and at the sub-segment level.

4.9 DATA QUALITY

The contribution of each dataset is limited by the quality of the dataset. The quality of the dataset has several components outlined in the following sections. A system can only provide an output or outcome that is only as successful as the input data. Thus, it is useful to be aware of these data quality components, and to measure them and use the measurement to prioritize where improvements can be made in addition to measuring the performance of the overall system.

4.9.1 ACCURACY

Data accuracy refers to the correctness, or effort of the values of the dataset compared to the actual phenomenon measured or represented. There are several factors that contribute to the accuracy of detection data, including calibration, maintenance, and environmental conditions. Each of these factors has a different impact on different sensor and data source types. Passive observation sensors are generally more impacted than probe detection systems receiving a signal transmitted from vehicles.

Ground truth data refers to direct observation that is considered to be 100 percent accurate. This is typically costly to obtain at scale, and used to validate more efficient detection methods or to train or test a model. For example, video can be recorded and vehicles can be counted by an observer. This can be used to compare against relative volumes obtained from probe detection systems to calibrate a coefficient or other imputation method to produce a volume value from a relative volume value.

4.9.2 AVAILABILITY AND COVERAGE

Data availability can be defined as the number of measurement values available over the number of expected measurement values. Spatial availability deals with locations where measurement values are available. Places where there is a lack of instrumentation, or where there is either no communication or otherwise no means for collecting the data can contribute to gaps in spatial coverage. Temporal coverage deals with the time intervals where measurement values are available. System and communications can have temporary unavailability.

The data coverage of a particular dataset can be expressed as a percentage over a time dimension or over a set of sensors e.g. a spatial region. This defines more than one dimension that can be observed, and this can be presented in different ways. For example, one report of the availability for a sensor or group of sensors could define a temporal extent beginning with the earliest record, and ending with the most recent record, and could calculate the percentage of time intervals where a record is expected and is present. Another report could specify a particular time interval, such as the current, or most recent interval expected; and output the percentage of sensors that have a record present for that interval in the dataset. These two reports could be combined for a two-dimensional heat map, allowing the user to quickly see where there may be issues in data availability for the dataset. Data coverage is similar to availability, but refers to the set of reporting locations, typically spatial in nature, over which data is expected to be reported on. This can refer to the extent of the coverage, or the precision of the spacing interval of the coverage.

4.9.3 VALIDATION

Data validation involves verifying that data is accurate, complete, consistent, and adheres to predetermined rules or criteria. It is an essential step in data quality management and is performed to identify and correct errors or anomalies in the data before it is used for analysis, reporting, or decision-making. Here are some common techniques and practices involved in data validation:

1. *Data Integrity Checks:* These checks verify the integrity of the data, ensuring that it is not corrupted or altered during storage or transmission. It can involve techniques such as checksums, hash functions, or digital signatures.
2. *Format and Type Validation:* This involves validating that the data is in the correct format and conforms to the expected data types. For example, checking that a date field follows a specific format or ensuring that numeric fields contain only numerical values.
3. *Range and Constraint Validation:* Validating data against predefined ranges or constraints ensures that it falls within acceptable limits. For instance, checking that volume values are within a specific range and do not exceed the highway capacity manual for maximum volume that can be excepted for a lane or a turning movement.
4. *Cross-Field Validation:* This involves checking the consistency and coherence of data across multiple fields. For example, if occupancy is high, then speed cannot also be high; and if volume is low, so will occupancy likely be.
5. *Completeness Validation:* This ensures that all required data fields are present and populated. It involves checking for missing values or null entries in critical fields.
6. *Duplicate Check:* This involves identifying and handling duplicate records or entries within the dataset. Duplicate checks can be performed based on one or multiple fields, and the appropriate actions, such as merging or removing duplicates, can be taken.

4.9.4 DATA FORMAT: MACHINE ACCESSIBILITY AND READABILITY

One consideration for data quality is the ability to access and interpret the information. For algorithms, having the information in a consistent, unambiguous, machine-readable format is critical for usability. Defining protocol standards and data schemas describes a consistent structure of the information and a consuming system can validate the incoming data. Even still, schemas can leave interpretation if they do not specify in sufficient detail to define how to express the information, e.g., a free text field without any further specification as to what goes into the string value. However, if too rigid, the schema can overly constrain the expression of information to now allow elements to be expressed such as an enumeration that only includes a few allowed values that does not cover all possibilities that could be encountered, such as defining a set of movement restrictions without a day of week or time range over when the movement restriction applies. One aspect that adds difficulty is when there are multiple manufacturers of an information-producing system of a certain type. The standard attempts to help by allowing a consumer to be presented a consistent information format across different systems producing information; however, each system may have nuances that are not able to be represented by the standard.

For example, probe detection systems depend on capturing input from vehicles; however, not all vehicles are equipped or configured to participate in the system. Connected vehicle on-board units are not deployed in all vehicles. Bluetooth devices are not in all vehicles, and other vehicles may have multiple Bluetooth devices – both impacting the ability to get an accurate count of vehicles. 5% to 15% is generally considered a minimum capture rate required to get reasonably accurate travel time and speed conditions. Volumes cannot be measured directly with such a low capture rate; however, relative volumes can be useful for imputing volumes if combined with historic ground truth and the relative volumes during that time to calibrate a bias or coefficient.

4.9.5 DATA LATENCY

Sensor data latency refers to the time delay between when a sensor collects data and when that data becomes available for processing or analysis. It is a critical factor in many applications where real-time or near-real-time data is required. Acceptable levels of latency can vary depending on the specific application and use case. Some applications require extremely low latency (milliseconds to microseconds), while others may tolerate slightly higher latency (seconds to minutes). Here are some key points regarding sensor data latency:

1. *Sensor Technology:* Different types of sensors have varying levels of latency. For example, a simple temperature sensor may provide near-instantaneous data, while a complex imaging sensor or a remote environmental monitoring sensor may have longer latency due to processing or transmission requirements.

2. *Data Transmission:* Latency can be introduced during the transmission of sensor data from the sensor location to the data processing or storage infrastructure. This is particularly relevant in wireless or networked sensor systems where data may need to be transmitted over long distances or through networks with varying bandwidth or congestion levels.
3. *Communication Protocols:* The choice of communication protocols used for transmitting sensor data can impact latency. For example, protocols with higher overhead or encryption requirements may introduce additional delays compared to more lightweight protocols.
4. *Processing and Aggregation:* Depending on the complexity of the sensor data and the processing infrastructure, additional latency may occur during data processing and aggregation. This can involve tasks such as data filtering, normalization, or feature extraction, which can introduce delays before the data is ready for analysis.

4.10 CONCLUSIONS

This chapter has provided a comprehensive exploration of the diverse array of traffic data modalities essential for effective transportation management. Through the elucidation of fixed point detection technologies, including loop detectors, microwave detectors, and advanced video and LiDAR systems, we have gained insights into the real-time monitoring of traffic dynamics at specific locations.

Furthermore, the examination of vehicle probe detection and trajectory data has highlighted the dynamic and granular nature of data collection, enabling a nuanced understanding of individual vehicle behaviors within transportation networks. Incident data management protocols, coupled with insights into supplementary datasets such as third-party traffic conditions and multimode data, underscore the multifaceted approach required for comprehensive traffic analysis and management.

Moreover, the discussion on automated traffic signal performance measures and corridor-wide performance metrics has emphasized the importance of continuous monitoring and evaluation to optimize traffic flow and enhance system efficiency. By incorporating data-driven insights into long-range planning, medium-scale planning, and active operational decision-making, transportation agencies can better anticipate and respond to evolving traffic demands and challenges.

Finally, the chapter has underscored the paramount importance of data quality, emphasizing the need for accuracy, availability, validation, and machine accessibility/readability. By ensuring the integrity and reliability of traffic data, stakeholders can make informed decisions and implement targeted interventions to improve traffic management strategies and ultimately enhance the overall safety, efficiency, and sustainability of transportation systems.

5 Data Mining and Machine Learning

Abstract - This chapter briefly explores the critical areas of clustering, outlier detection, and neural networks within the context of machine learning and its application to traffic data analysis. The chapter delves into various algorithms, methodologies, and applications of these concepts, with a special focus on traffic data analysis, including traffic state prediction, intersection performance, and detection of traffic interruptions. By giving a brief introduction to machine learning techniques, this chapter aims to provide insights into how these can be leveraged for applications in data driven management of traffic networks.

The advent of machine learning has revolutionized data analysis, pattern recognition, and predictive modeling across various domains, including transportation and traffic management. Among the myriad of techniques, clustering, outlier detection, and neural networks stand out for their ability to uncover patterns, identify anomalies, and predict outcomes from complex datasets. This chapter presents a brief explanation of these techniques, emphasizing their application to traffic data analysis and the unique challenges and opportunities they present.

Clustering, a foundational unsupervised learning technique, is pivotal in identifying natural groupings in data, enabling meaningful segmentation and analysis without prior labeling. This chapter covers key clustering algorithms such as K-Means and hierarchical clustering, alongside distance metrics and methods to assess clustering quality. The discussion extends to the application of clustering in summarizing traffic intersection performance, highlighting its utility in managing vast amounts of traffic signal data.

Outlier detection, or anomaly detection, plays a crucial role in ensuring data integrity and enhancing model performance by identifying data points that significantly deviate from the norm. This section explores both supervised and unsupervised approaches to outlier detection, with a focus on its importance in detecting traffic interruptions and improving road network management.

Neural networks, with their ability to model complex relationships through layers of interconnected units, have seen widespread adoption in processing traffic data. From basic feed-forward networks to advanced architectures like recurrent neural networks (RNNs), long short-term memory networks (LSTM) , and transformers, this chapter examines how these models can predict traffic states, optimize signal timing, and contribute to effective traffic management solutions.

5.1 CLUSTERING

Clustering is an unsupervised learning technique that involves the grouping of similar data points into clusters, based on their inherent patterns and similarities. Unlike supervised learning, clustering does not rely on labeled data but instead identifies natural structures within the data itself. It has numerous applications in various domains, including data analysis, pattern recognition, image segmentation, and customer segmentation.

There are several types of clustering algorithms, but below are some of the most commonly used methods.

- K-Means stands as a widely used partition-based clustering algorithm, which segments data into 'k' clusters—where 'k' represents a predetermined number set by the user. The algorithm iteratively assigns each data point to the nearest cluster centroid and adjusts the centroids until convergence.
- Hierarchical clustering constructs a dendrogram, forming a tree-like arrangement of nested clusters. Unlike other methods, it doesn't necessitate the user to predefine the number of clusters. There are two primary approaches to hierarchical clustering: agglomerative, which initially treats each data point as an individual cluster and progressively merges them, and divisive, which begins with all data points in one cluster and iteratively divides them.

Distance metrics play a crucial role in clustering algorithms as they determine the similarity or dissimilarity between data points. Common distance metrics used in clustering include the following:

- Euclidean Distance - The most widely used distance metric, it calculates the straight-line distance between two points in Euclidean space.
- Manhattan Distance - Also known as L1 norm, it calculates the sum of absolute differences along each dimension between two data points.
- Cosine Similarity - Measures the cosine of the angle between two vectors, indicating their similarity regardless of their magnitude.

Assessing the quality of clustering results is essential to understanding the effectiveness of the algorithm. However, since clustering is an unsupervised task, there are no definitive labels to evaluate against. The following are two unsupervised scoring methodologies.

- Silhouette Score: Measures how well-separated the clusters are and assigns a score between -1 to 1 for each data point. Higher scores indicate well-defined clusters.
- Davies-Bouldin Index: Measures the average similarity between each cluster and its most similar cluster. Lower values suggest better clustering.
- Calinski-Harabasz Index: Measures the ratio of the between-cluster dispersion to the within-cluster dispersion. Higher values indicate better-defined clusters.

Selecting the appropriate number of clusters is a pivotal stage in clustering algorithms. The optimal value ensures that the clusters accurately reflect the underlying structure within the data, rendering them meaningful and representative. Selecting an inappropriate number of clusters can lead to either overly granular or overly broad clusters, undermining the effectiveness of the clustering process. The Elbow Method is a technique used to identify the optimal number of clusters based on the variance explained as a function of the number of clusters. It involves plotting the within-cluster sum of squares (WCSS) or the sum of squared distances from each point to its assigned cluster centroid for different values of the number of clusters. The "elbow" point on the plot, where the WCSS starts to level off, indicates the optimal number of clusters. The idea is to choose the number of clusters where adding more clusters does not significantly reduce the WCSS. Some other methods include techniques based on silhouette scores, gap statistics etc.

Density-based clustering algorithms like DBSCAN (Density-Based Spatial Clustering of Applications with Noise) eliminate the need for users to specify the number of clusters in advance. Instead, they group data points based on local density. Points that belong to dense regions are assigned to clusters, while points in sparse regions are treated as noise or outliers.

Clustering of spatio-temporal data presents unique challenges and opportunities, as it involves understanding patterns and trends in both space and time. Spatiotemporal data is prevalent in various fields, including environmental science, transportation, epidemiology, and urban planning. Spatiotemporal data points that are geographically close may also have a temporal dependency, leading to spatial clusters evolving over time. Dividing the spatiotemporal domain into grids allows for efficient clustering by considering both spatial cells and time intervals. By leveraging specialized clustering algorithms and visualization techniques, one can analyze the complex interplay between space and time in various real-world systems.

The newest generation of traffic signal controllers can record both signal phasing information and vehicle arrivals and departures at a very high resolution (10 Hz).This has resulted in the development of many new measures of effectiveness (MoEs) including the development of the automated traffic signal performance measures (ATSPM)[1] and other related efforts. This allows traffic engineers to quantitatively analyze the performance of signalized intersections. But this vast amount of information without a birds eye view of whats happening at an intersection or corridor level adds a burden to transportation professional. To present this information in succinct and actionable form, several clustering techniques can be leveraged for such purposes. To this extent, how clustering can be leveraged to summarize intersection performance on a corridor for various times of the day and days of the week is presented in detail in chapter 8

[1]`https://udottraffic.utah.gov/atspm/`

5.2 OUTLIER DETECTION

Outlier detection, also known as anomaly detection, is a critical task in data analysis and machine learning. Outliers are data points that significantly deviate from the majority of the dataset or exhibit unusual patterns compared to the rest of the data. Identifying and handling outliers is essential for maintaining data quality, improving model performance, and gaining valuable insights into abnormal events or errors. Univariate outliers are data points that have extreme values along a single feature or attribute. Multivariate outliers are data points that have extreme values when considering multiple features or attributes simultaneously. Detecting multivariate outliers requires analyzing the correlation and interaction between different features.

Outlier detection is valuable across various domains, including fraud detection, network intrusion detection, fault detection, and healthcare monitoring. Several machine learning techniques can be employed for anomaly detection, each with its strengths and limitations. They can be broadly classified into supervised and unsupervised learning algorithms.

Outlier detection in a supervised setting can be formulated as a binary classification problem. In this approach, the goal is to build a model that can distinguish between normal (inlier) data points and anomalous (outlier) data points. The labeled data is used to train the classifier, where the normal data points are assigned the positive class (e.g., label 1), and the anomalous data points are assigned the negative class (e.g., label 0). Once the classifier is trained, it can be used to predict whether new, unseen data points are normal or anomalous. By casting anomaly detection as a classification problem, we can leverage the rich set of tools and algorithms available for classification tasks. It allows for more straightforward integration of anomaly detection into existing machine learning pipelines and frameworks.

Anomaly detection as an unsupervised learning algorithm involves identifying anomalies or outliers in a dataset without using labeled data. In this approach, the algorithm is trained on the dataset containing only normal data points, and it learns to model the normal behavior of the data. Once trained, the algorithm can then identify data points that deviate significantly from the learned normal behavior as potential anomalies. Unsupervised anomaly detection is valuable when labeled anomaly data is scarce or not available. In clustering-based methods, outliers can be detected as data points that do not belong to any cluster or belong to small clusters. Algorithms based on autoencoders identify anomalies as data points that are farthest or difficult to reconstruct accurately.

Outlier/Anomaly detection techniques find several applications in transportation. For example, detecting traffic interruptions is critical aspect in managing road networks. These interruptions/incidents may arise due to vehicle breakdowns, accidents, debris etc. Outlier detection approaches can be leveraged to detect traffic interruptions and thus mitigate adverse affects such as congestion. A data driven framework for detecting such traffic interruptions is detailed in chapter 9

5.3 NEURAL NETWORKS

Artificial neural networks are a class of machine learning models inspired by biological neural networks. They consist of systems of interconnected units, with each unit (called "neurons") taking multiple inputs and emitting a single output, based on an activation function. Neurons are connected to each other in layers and are trained on a dataset via the back-propagation rule.

With the advent of powerful computing technologies such as graphics processing units (GPUs) and tensor processing units (TPUs) and economical large cloud storage technologies, large multilayer neural networks (with thousands to millions of tunable parameters) are being trained on equally large datasets running into 100s of gigabytes. This paradigm is referred to as "deep learning." Deep learning can be understood as a representation-learning method that consists of neural networks that learn a hierarchy of representations, ranging from simple features to more abstract ones.

5.3.1 FEED FORWARD NEURAL NETWORKS

Feed Forward Neural Networks (FNNs), commonly known as Artificial Neural Networks (ANNs), are a fundamental class of deep learning models used for various tasks, including classification, regression, and pattern recognition. Inspired by the structure of the human brain, FNNs consist of interconnected artificial neurons (nodes) organized in layers. Information flows forward through the network from the input layer to the output layer, without any cycles or feedback connections.

A typical FNN consists of three main layers.

Input Layer: The first layer of the network that receives the raw input data. Each node in the input layer represents a feature or attribute of the data.

Hidden Layers: One or more layers sandwiched between the input and output layers. These layers contain hidden neurons and are responsible for learning complex patterns and features from the input data.

Output Layer: The final layer that produces the network's output. The number of nodes in the output layer depends on the specific task. For binary classification, there is one node for the binary output (e.g., 0 or 1), while multi-class classification tasks may have multiple output nodes, one for each class.

Feed Forward Neural Networks get their name from the forward flow of information during inference, known as forward propagation. The input data passes through the hidden layers, where each neuron computes a weighted sum of its inputs, applies an activation function, and passes the result to the next layer. The activation function introduces non-linearity to the network, enabling it to learn complex relationships in the data.

Activation functions play a crucial role in FNNs by introducing non-linearities, enabling the network to model complex relationships in the data. Commonly used activation functions include Sigmoid, Tanh, ReLU (Rectified Linear Unit), Leaky ReLU, and Softmax (for multi-class classification).

FNNs face challenges like overfitting, vanishing gradients, and hyperparameter tuning. Advancements in deep learning, such as the introduction of deeper architectures (Deep Neural Networks), regularization techniques (Dropout, Batch Normalization), and optimization algorithms (Adam, RMSprop), have significantly improved the performance and scalability of FNNs.

5.3.2 RECURRENT NEURAL NETWORKS

Recurrent neural networks (RNN) are a class of artificial neural network, which is especially well-suited for modeling temporal sequences. These networks process input sequences within the context of their internal hidden state ("memory") to arrive at the output. The internal hidden state is an abstract representation of previously seen inputs. Thus, they are capable of dynamic contextual behavior.

During training, backpropagation through time (BPTT) is used to compute the gradients and update the network's parameters. BPTT extends backpropagation to the recurrent connections, considering the entire sequence of data during training. However, RNNs can suffer from vanishing or exploding gradients, which can make training challenging for long sequences. This limitation led to the development of more advanced RNN variants like long short-term memory networks (LSTM) and Gated Recurrent Units (GRU).

LSTM and GRU are advanced RNN architectures designed to overcome the vanishing gradient problem. They introduce gating mechanisms that allow the network to selectively remember or forget information in the hidden state. These gating mechanisms improve the flow of information across time steps and enable RNNs to capture long-range dependencies in the data effectively. LSTM network consists of the following key building blocks

- Memory Cell: The core of LSTM is the memory cell, which stores and updates information over time. The memory cell's contents are regulated by three main gates:
- Forget Gate: The forget gate determines what information to discard from the previous hidden state. It takes the previous hidden state and the current input as inputs and outputs a value between 0 and 1 for each element in the memory cell. A value close to 0 indicates that the information should be forgotten, while a value close to 1 indicates that it should be retained.
- Input Gate: The input gate controls which new information should be stored in the memory cell. It takes the previous hidden state and the current input as inputs and outputs a value between 0 and 1 for each element in the memory cell. A value close to 0 means that the information should not be stored, while a value close to 1 means that it should be stored.
- Output Gate: The output gate determines what information to output from the memory cell to the current hidden state. It takes the previous hidden state and the current input as inputs and outputs a value between 0 and 1 for each element in the memory cell.

The gates in LSTM enable the network to learn long-range dependencies and remember important information over extended sequences.

GRU combines the forget and input gates of LSTM into a single update gate and eliminates the output gate, resulting in a more streamlined architecture. GRU consists of the following key building blocks.

- Update Gate: The update gate combines the functionality of the forget and input gates in LSTM. It takes the previous hidden state and the current input as inputs and outputs a value between 0 and 1 for each element in the memory cell. A value close to 0 indicates that the information should be forgotten, while a value close to 1 indicates that it should be retained.
- Reset Gate: GRU introduces an additional gate called the reset gate, which controls how much of the previous hidden state should be forgotten. It takes the previous hidden state and the current input as inputs and outputs a value between 0 and 1 for each element in the memory cell. A value close to 0 means that the corresponding element in the previous hidden state should be forgotten, while a value close to 1 means that it should be retained.

GRU simplifies the LSTM architecture by combining some of the gates and has been shown to perform comparably to LSTM while requiring fewer parameters and being computationally more efficient.

Teacher forcing [149] is a common training technique for training RNNs (and thus GRUs). In teacher forcing, the actual answer of the previous time step is provided to the RNN while it predicts the current output. An error in the previous output could cause a large error in the current output, which in turn would accumulate over time steps. Teacher forcing remedies this by penalizing the network for the wrong answer at that time step but doesn't allow the network to commit a series of errors based on the initial error. This technique has been shown to lead to faster convergence. Given our large dataset and models, we employ this technique to speed up model training.

5.4 TRANSFORMERS

The Transformer architecture, introduced in the landmark paper "Attention Is All You Need" by Vaswani et al. [138], marked a significant breakthrough in natural language processing (NLP) and other sequence-to-sequence tasks. Transformers are a class of deep learning models that rely on attention mechanisms, enabling them to process sequential data with remarkable efficiency and effectiveness. Unlike traditional recurrent neural networks (RNNs) and convolutional neural networks (CNNs), Transformers are not limited by sequential processing, making them ideal for tasks where long-range dependencies are crucial.

The Transformer architecture consists of an encoder-decoder framework. In the context of NLP, the encoder processes the input text, and the decoder generates the output text. Each encoder and decoder layer are composed of multiple self-attention and feed-forward neural networks.

The core of the Transformer architecture is the attention mechanism, which allows the model to focus on specific parts of the input sequence when processing each element. Attention mechanism [138] is a deep learning technique for effectively dealing with long-range dependencies in neural models. The broad idea is to create linkages between the current context vector (which, in the case of GRUs, would include the last hidden state) and the entire source input (or its abstract representation). Thus, the context field of the model is enhanced and is no longer prone to forget events in the distant past.

In self-attention, a sequence is represented as three vectors: Query, Key, and Value. Each element in a sequence, such as a word in a sentence, is associated with three vectors: the query vector, the key vector, and the value vector. These vectors capture different aspects of the input data. The query vector represents the element being considered, the key vector holds information about other elements in the sequence, and the value vector encodes the information that the model should focus on or retrieve from the associated key. The self-attention mechanism computes attention scores between query and key vectors to determine the relevance or importance of other elements in the sequence.

The attention weights are computed as the dot product between the Query and Key vectors, scaled by a factor to control the softmax distribution. A softmax distribution is a probability distribution commonly used in machine learning and deep learning to convert a vector of arbitrary real numbers into a probability distribution. It takes an input vector and normalizes it into a set of values that sum to 1, making it suitable for tasks where you want to assign probabilities to multiple classes or options.

The values are then weighted by these attention weights and summed up to obtain the context vector for each position in the sequence. To capture different types of relationships in the data, the Transformer uses multiple attention heads in parallel. Each attention head learns a unique attention pattern, enhancing the model's ability to process diverse dependencies and patterns in the data.

Since Transformers lack the inherent order of sequential data like RNNs, positional encodings are added to the input embeddings to provide positional information to the model. These encodings are designed to be learnable and effectively encode the order of elements in the sequence. Transformers have revolutionized the field of deep learning, enabling more powerful and efficient models that can handle complex tasks. Their ability to capture long-range dependencies and context has made them a go-to choose for various applications.

Deep learning networks have been successfully applied to traffic data for traffic state prediction [142] for short-term (5-30 minutes), medium-term (30-60 minutes) and long-term (1+ hour) time windows. These models can learn the complex intersection dynamics and interplay between different factors such as vehicle arrivals, signal timing information, driving behaviour, upstream intersection state etc. In chapters 12, 13 we describe how these neural network models can be leveraged to modelling input and output flow dynamics between a pair of intersections, measures of effectiveness such as wait times under different signal timing plans and their potential use in signal timing optimization software.

5.5 CONCLUSIONS

In conclusion, this chapter has traversed the landscape of clustering, outlier detection, and neural networks, shedding light on their pivotal role and practical application within the domain of traffic data analysis. Through a nuanced exploration of diverse algorithms and methodologies, we have unveiled the power of these machine learning techniques in unraveling patterns, identifying anomalies, and forecasting outcomes, thereby fostering the development of more agile and responsive traffic management systems.

Clustering has emerged as a cornerstone of unsupervised learning, enabling the identification of natural groupings within data without prior labeling. By delving into clustering algorithms such as K-Means and hierarchical clustering, alongside essential distance metrics, we have elucidated its application in summarizing traffic intersection performance and managing vast troves of traffic signal data.

Furthermore, outlier detection has been spotlighted as a critical component in maintaining data integrity and enhancing model efficacy by pinpointing data points that deviate significantly from the norm. Whether through supervised or unsupervised approaches, the detection of anomalies plays a pivotal role in mitigating disruptions and optimizing road network management.

Moreover, the chapter has navigated the landscape of neural networks, from basic feed-forward architectures to advanced models like recurrent neural networks (RNNs) and transformers. By harnessing the power of neural networks, traffic engineers can delve into intricate traffic dynamics, predict traffic states, optimize signal timing, and ultimately enhance the efficiency of traffic management solutions.

In summation, this chapter underscores the transformative potential of clustering, outlier detection, and neural networks in reshaping the landscape of traffic data analysis. By harnessing these machine learning techniques, transportation stakeholders can unlock deeper insights, foster more adaptive strategies, and pave the way for safer, more efficient, and more sustainable transportation networks of the future.

6 Traffic Simulation Frameworks for Data Generation

Abstract: In this chapter, we discuss the role of traffic simulation frameworks. We describe different types of simulation frameworks, based on the complexity of modeling. We describe some important traffic simulation frameworks, that are often used in conjunction with machine learning algorithms, either as sources of training data, or as environments of reinforcement learning algorithms. We also discuss how a traffic simulation can be calibrated based on real-world considerations.

With advances in electronics, sensing, computing, data storage and communications technologies, Intelligent Transportation Systems (ITS) help traffic engineering practitioners by augmenting their experience with data-driven insights.

However, effectively and efficiently modeling an urban traffic network, is a challenging task. Not only have the behavior and attributes of the static elements of the network (such as road links, number of lanes, speed limits, etc.) have to be accounted for, but even the dynamic elements such as actuated signals and vehicles, have to be accurately represented. Analytical approaches are often limited in their ability to capture the complex nature of these interactions [15, 118]. Hence researchers in the field of traffic engineering have turned to specialized softwares [15] designed to model, simulate and visualize traffic phenomena. These traffic simulation frameworks serve a variety of functions including:

- Replicating and analyzing complex traffic phenomena using real-world traffic sensor data and basemaps. This allows traffic engineers and authorities to study the traffic scenario in detail.
- Formulating and evaluating new traffic management strategies [146] such as signal timing plans etc. for improving traffic Measures of Effectiveness (MoEs) [143]. This allows for exploring counterfactual "what-if" traffic scenarios without adversely interfering with real-world traffic flows.
- Using simulators to capture data for training machine learning models [8], by recording data for Supervised/Unsupervised Learning and as dynamic environments for reinforcement Learning [126]. These models can then be deployed for a variety of tasks such as MoE estimation, traffic signal management, autonomous driving, [48] etc.

With these benefits, traffic simulation frameworks are an indispensable tool in the field of traffic engineering but also in the field of machine learning. However, one of the key concerns of traffic simulation is the computational cost. Microscopic traffic simulators especially are computationally-expensive [105] as they model the interactions between the agents (such as vehicles, traffic signals, etc.) at fine spatial (down to centimeters) and temporal resolutions (down to a tenth of a second). One important aspect of traffic simulation is calibration. [42, 43] explore the sensitivity analysis of computationally-expensive traffic simulation models. The number of simulations needed to cover a multidimensional input parameter space (such as car-following and lane-changing parameters) for calibration grows exponentially with the number of such inputs.

One of the first major works applying parallel computing was PARAMICS [11] (Parallel microscopic simulation of road traffic) which in 1996 was developed with a 256-node CRAY T3D as the target machine. It was able to simulate approximately 200,000 vehicles on 20,000 miles of roadway. It has since then been updated and commercially marketed as a proprietary licensed software. FastTrans [130] is a parallel, distributed-memory simulator using a queue-based event-driven approach to traffic microsimulation. A parallel implementation of the Transportation Analysis and Simulation System (TRANSIMS) traffic microsimulation is described in [101]. It uses domain decomposition, wherein each CPU is responsible for a different geographical area of the simulated region. [5] presents the results of the parallelization of the AIMSUN2 simulator. Microscopic simulator SUMO uses multi-threading parallel computation for vehicle routing but the remaining parts of SUMO are single-core. MATSim which is an open-source framework for large-scale agent-based transport simulations, is parallelized by the BEAM [121] (Behavior, Energy, Autonomy, and Mobility) framework. [47] presents OTM-MPI, which extends the Open Traffic Models platform (OTM) for running parallelized macroscopic traffic simulations. [74] introduces ParamGrid, a scalable and synchronized framework that distributes the PARAMICS across a cluster of commodity machines. [113] describes SMARTS (Scalable Microscopic Adaptive Road Traffic Simulator) a distributed microscopic traffic simulator that can utilize multiple independent processes in parallel. CityFlow [162] is a recent open-source traffic simulator, that claims to be faster than SUMO, by using multi-threading, optimized data structures and efficient algorithms. It has been developed with a focus for deep reinforcement learning applications in traffic signal management. [20] introduces a parallel, congestion-optimized version of SUMO. It breaks up the simulation map using METIS graph partitioning tool and also reduces granularity of updates during known congested situations.

Also, traffic co-simulation is becoming popular these days. [12] investigates the use of CARLA, SUMO/VISSIM for traffic co-simulation, Carsim or MATLAB/Simulink for vehicle dynamics co-simulation and Autoware for autonomous driving algorithm co-simulation. OpenCDA [157] is an open Cooperative Driving Automation (CDA) Framework with SUMO controlling the general traffic and CARLA controlling the co-operative CAVs (Connected and Autonomous Vehicles). [87] combines road traffic simulator SUMO and network simulator NS3 for VANET (Vehicular Ad hoc Network) applications. Similarly VEINS [122] combines SUMO and OMNet++. [50] couples open-source MATSim and SUMO traffic simulators to handle mesoscopic and microscopic simulations respectively.

6.1 TRAFFIC SIMULATION FRAMEWORKS

Traffic simulation frameworks [136, 32] are computational implementations of traffic models with dynamic components (vehicles, traffic signals, pedestrians etc.) and static components (road geometry and linkages etc.). A traffic simulation consists of a traffic scenario which consists of a basemap that defines the static components such as the topology of roads with lanes, junctions which connect these roads etc. These static components usually do not change their behaviors in the short term (i.e. in seconds or minutes). On this basemap, dynamic components usually change their states based on predefined behaviors (i.e. cars will change their locations based on their speeds and accelerations, traffic signals will change their light configuration based on signal plan etc.). The simulation is started and is allowed to evolve in time. We can thus simulate a variety of basemaps and behaviors, and estimate different measures of effectiveness (such as queue lengths and travel times).

Based on the level of detail presented to the user, we can broadly classify traffic simulators as:

1. Macroscopic Simulators: These simulators simulate high-level flow characteristics such as speed, density, etc. by considering the aggregate behaviors of a large collection of vehicles (akin to fluid dynamics). While these simulators allow for computationally-inexpensive simulations of large traffic grids, they lack the capability for fine-grained space-time analysis of traffic behaviors.
2. Microscopic Simulators: These simulators simulate fine-grained vehicle behaviors, often at the individual level. Some may simulate the roadway divided into cells which can be occupied or not occupied by a vehicle, and these cells progressively switch on and off, to indicate vehicle movement. Another approach simulates individual vehicles as independent agents that move over roadways, and have pre-programmed behaviors such as a car-following model, behavior at intersections etc. These simulators are usually computationally-expensive, and are usually used to analyze an intersection or a corridor.
3. Mesoscopic Simulators: These simulators strike a balance between Microscopic and Macroscopic simulators. They usually analyze groups of vehicles, such as homogeneous platoons.

In this study, we focus on microscopic simulators, as they are the most computationally-expensive. Below, we describe three important microscopic traffic simulation softwares, VISSIM, SUMO and CARLA. Other microscopic traffic simulators currently in use include CORSIM [51], AIMSUN [13], PARAMICS [127] etc.

6.1.1 VISSIM

VISSIM [89] is a microscopic multimodal traffic flow simulation package developed by PTV Planung Transport Verkehr AG, Germany. VISSIM is widely used, with over 2,500 cities worldwide using it.

VISSIM supports a large variety of road types, intersections and signals, and simulates a variety of vehicles such as cars, trucks, public transport, etc. VISSIM also supports pedestrian dynamics as well. VISSIM is made for the Microsoft Windows Operating System and is a proprietary closed-source software with paid licenses. These licenses restrict the number of machines that the software can be run on. VISSIM was first developed in 1992 and is actively updated. VISSIM supports a component object model (COM) programming interface. This interface allows users to develop and implement their own code that can use a VISSIM network and simulation, as well as modify it. This includes network topology, vehicles and flows, traffic light behavior, etc. Supported programming languages include Python, C++, and Visual Basic.

Retime: VISSIM[1] is a Software-as-a-service (SAAS) platform that uses cloud computing to run multiple instances of VISSIM for optimization of traffic signal timings. The user uploads a VISSIM network file. A Genetic Algorithm (GA) is used to evaluate and find the best signal timings. Results and analytics for various runs are visually presented via a web interface. Both VISSIM and Retime are commercial proprietary closed-source frameworks.

6.1.2 SUMO

Simulation of Urban MObility (SUMO) [88], developed by the Institute of Transportation Systems of DLR (German Aerospace Center) and the National Aeronautics and Space Research Center of the Federal Republic of Germany in Berlin, is a software tool widely used for road traffic simulation. SUMO, licensed under the GPL, is an open-source, highly portable, microscopic, and continuous traffic simulation package designed for managing large road networks. It utilizes its own file formats for traffic networks but is capable of importing files encoded in popular formats like OpenStreetMap [104], VISSIM [89]. Implemented in C++ and utilizing only portable libraries, SUMO is lightweight and efficient. SUMO is single-threaded i.e. uses only one CPU core but several parallel SUMO processes can be spawned allowing for parallel simulations.

SUMO enables one to import or automatically generate signal timing plans for traffic lights. It supports various phase sequences and durations. Additionally, real-world maps can be imported using OpenStreetMap.

The SUMO package consists of a suite of several different interrelated programs:

- SUMO: The heart of the package. It is a microscopic simulation engine with no visualization.
- SUMO-GUI: A visualization tool for SUMO, shows the map and the vehicles moving on it. Also, it has a graphical user interface to interact with SUMO.
- NETCONVERT: Network importer and generator. It reads road networks in different formats and converts them into the native SUMO-format

[1]http://retime.online/aboutRetime.html

- NETEDIT: A graphical network editor that allows for the editing of maps by hand.
- DUAROUTER: Computes fastest routes through the network and performs the DUA (Dynamic User Assignment)
- JTRROUTER: Computes routes using given junction turning percentages
- DFROUTER: Computes routes based on given induction loop detector measurements
- OD2TRIPS: Decomposes O/D-matrices into single-vehicle trips
- TraCI is a package that ships with SUMO and is short for "Traffic Control Interface". It gives access to a user-written Python code (or an external Python program) to a road traffic simulation running in SUMO. It allows for the retrieval of values of simulated objects and the manipulation of their behavior while the simulation is running. TraCI uses a TCP-based client/server architecture to provide access to SUMO.

6.1.3 CARLA

CARLA [33] is an open-source driving simulator, built primarily for autonomous driving research. It is based on the Unreal graphics engine, used by high-end video game developers. CARLA uses a scalable client-server architecture. The server runs the simulation, while client modules can control the logic of actors such as vehicles. CARLA provides open digital assets such as urban layouts, buildings, vehicles etc. that can be used freely. An important point of distinction between CARLA and the other traffic simulators such as SUMO and VISSIM, is that the focus of CARLA is on driving vehicles, as opposed to traffic management. Hence, CARLA focuses on realistic 3D rendering of scenes as seen from a vehicle, than simulating traffic scenarios as a whole. CARLA allows for the generation and collection of various sensor data modalities such as color vision, depth-sensing, RADAR, LiDAR, etc. which would be collected by vehicles. In order to have the best of both worlds, i.e. realistic vehicle-level data collection, and realistic traffic management, CARLA can perform co-simulation with both SUMO and VISSIM. A SUMO (or VISSIM) simulation can be run, and CARLA can render the same simulation (especially road geometry, vehicle movements, and traffic signal behaviors). This allows us to collect additional vehicle-level simulated sensor data such as vision, RADAR, LiDAR, etc.

6.2 SIMULATION CALIBRATION

Before using a simulator for the purposes of signal timing optimization, it is vital to calibrate the model. The default settings for the simulation model may not reflect the field conditions accurately. Thus, an optimal solution obtained in simulation may not be applicable to the field conditions.

Generally, simulator calibration is done manually. Field readings are obtained through observation either in-person or via sensors such as cameras, etc. These readings are used to change the default settings of the model.

In this work, we focus on calibrating two important aspects of the simulation model: vehicle flows and speeding behavior. Calibrating vehicle flows is important, as the number of vehicles arriving at various intersections and their turn-movements directly impact the optimal green splits at each intersection. Calibrating the speeding behavior is important, as it impacts the arrival profile at downstream intersections, and thus affects the signal timing offsets between the intersections. As we are optimizing signal timing plans across a corridor, we need to consider both at-intersection effects (i.e. vehicle flows and turn-movement counts) and between-intersection effects (i.e. speeding behavior). We perform these two aspects of calibration on a nine-intersection model of an urban arterial corridor during a weekday PM peak hour traffic scenario.

We now briefly discuss data modes that are available.

6.2.1 FLOW CALIBRATION

In order to program flows in the simulator, it is important to know where the simulated vehicles need to start and end their journeys. An important point to remember is that during corridor simulation, the same vehicles that depart one intersection, arrive at the next intersection. Thus, the origins and destinations chosen can't piece-wise satisfy each intersection individually; they must satisfy the observed loop detector flows at all intersections for that period of consideration.

While the loop detector data provides an accurate representation of the total number of vehicles passing through the system at various intersections, it does not provide the origins and destinations of those vehicles. On the other hand, WEJO [2] data provides trajectory information of connected vehicles within the system during the period of consideration. We leverage this information to infer the origin-destination matrix of vehicles, such that they broadly satisfy the observed loop detector flows, while adhering to the turn-movement counts observed in WEJO data at the various intersections. We assume that WEJO data provides an accurate representation of all other non-tracked vehicles in the system; that connected and non-connected vehicles broadly follow the same aggregate origin-destination behavior. It is reasonable to assume that the vehicles in the WEJO data have a similar distribution of origins and destinations as other vehicles.

We use an iterative algorithm to refine the O-D (Origin-Destination) matrix obtained from WEJO data, which is sparse (1-5%). Hence, a large number of O-D matrix entries will be 0. As an initial guess, this O-D matrix is used to generate individual vehicle trips. Flow count waveforms at the inflow road segments (i.e. the inflows at the two edge intersections, and the side street inputs) are available from the loop detector data. For each of these inflow actuations, we probabilistically choose an O-D pair, using the O-D matrix as a probability matrix. For a vehicle at a particular origin, all valid destinations are sampled based on their relative likelihoods. In this manner, the simulator can be given the time and O-D pairs for each vehicle to be inserted.

[2]www.wejo.com

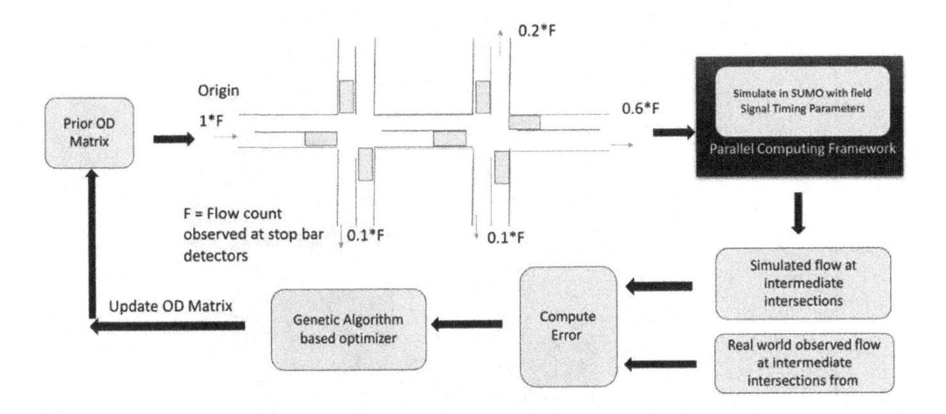

Figure 6.1: Probe trajectory data can be used to find maximum speeds attained by each vehicle in the dataset. These form a distribution when aggregated together. This can be used to set overspeeding parameters in the simulation.

Upon executing the simulation, simulated loop detector logs are recorded. These logs are used to compare the aggregate flows seen at the loop detectors in the simulation, with respect to the flows seen in the real-world. The error is computed, and the genetic algorithm-based optimizer chooses new potential O-D matrices as candidate solutions. The genetic algorithm then runs these new matrices in parallel and minimizes the computed error. The O-D matrix that minimizes the error at the loop detectors in simulation vs. the real-world, is chosen as the final O-D matrix.

6.2.2 OVERSPEEDING CALIBRATION

Another important aspect of simulator calibration is accounting for speeding behaviors. By default in SUMO, vehicles adhere to the posted speed limit. However, WEJO data analysis shows that there is significant speeding behavior, especially when vehicles move along the corridor just after the start of the green phase.

A significant fraction of the vehicles move at speeds well past the posted speed limit of 45 mph (72 kph) for some portions of their journey across their corridor. This distribution is used to calibrate the vehicle speeding behavior in SUMO. Vehicles will now speed beyond the posted speed limit as long as it is safe to do so, based on the above distribution.

This is done in SUMO using "speedFactor". This allows us to define a normal distribution with a mean and standard deviation, from which the vehicle speeding behavior will be sampled from. These parameters can be found using the WEJO data, and then fed into SUMO. The sparse probe trajectory data is used to calibrate the overspeeding behavior of vehicles in the simulation.

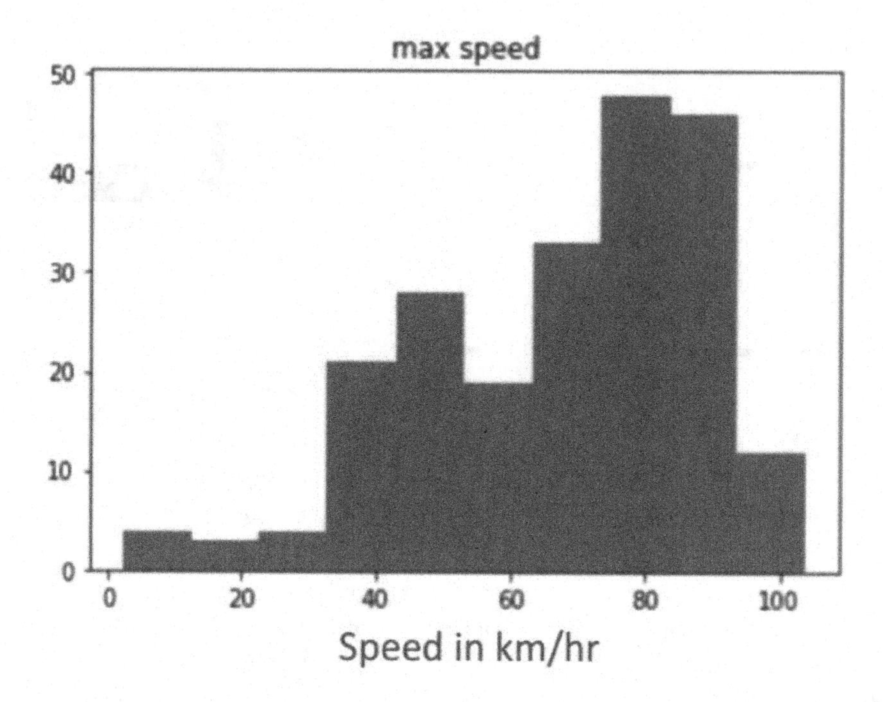

Figure 6.2: Flowchart showing the iterative refinement of origin-destination matrix. Genetic algorithm is used to improve the O-D matrix such that the loop detector aggregate flows at the various locations match the flows seen in the real-world.

6.3 DATASET GENERATION

In this section, we describe the data preprocessing and dataset generation pipeline. As an example, we generate a dataset for turning movement counts prediction. We shown how SUMO microscopic simulator programmed with real-world traffic flows and signal plans can be used to generate a dataset. Specifically, using SUMO helps us in the following ways:

- Datasets are often aggregated at 30-minute intervals, which is too coarse for our needs. SUMO allows us to generate data at 1-second resolution.
- Actual (ground truth) turning movement counts are not readily available nor is there a straightforward way of using controller log data for imputing them. SUMO logs allow us to get ground-truth turning movement counts.

One important aspect of simulations is modeling the demand to be as realistic as possible. Programming random flows or using coarse origin-destination matrices for generating flows may lead to unrealistic traffic distributions. The real-world dataset we use to program our simulation consists of controller log data from 329 signalized intersections in Seminole County, Greater Orlando Metropolitan Area. For traffic

Table 6.1
Actuated signal timing plan details.

Traffic Movement	Min Green (sec)	Max Green (sec)
Corridor-Through/Right	20	70
Side-Through/Right	10	30
Corridor Left	10	30
Side Left	10	30

flows, we use loop detector waveform data from advance detectors of these intersections and regenerate them in our simulations. We create a huge synthetic dataset of 30 million cycles, with traffic distributions similar to those observed in the real world.

Along with diverse but realistic traffic flows, we also need to program realistic traffic signal plans. In the real world, it is highly unlikely that a traffic authority would implement undesirable signal timing schemes for gathering data because that would have adverse real-world consequences. On the other hand, microscopic simulators offer us the flexibility of implementing undesirable signal timing plans. We study the Orlando dataset and derive a signal timing plan as shown in Table 6.1.

The signal timing plan for the intersection is an actuated signal timing plan with minimum and maximum times as shown in Table 6.1.

A yellow time of 5 seconds follows each phase. Thus, this leads to a theoretical minimum signal cycle length of 60 seconds and a maximum length of 180 seconds. This maximum is chosen with consideration of acceptable pedestrian wait times.

We use a three-stage approach for generating simulated data:

1. Generate a realistic intersection configuration in SUMO
2. Derive traffic waveforms from real data
3. Run parallel simulations using waveforms from step 2 and intersection configuration in step 1.

6.3.1 INTERSECTION CONFIGURATION

Our simulation consists of a one-intersection scenario with four approaches, based on standard NEMA (National Electrical Manufacturing Association) [133] phasing. It consists of four through or right movements and four left-turn movements, one of each movement for the four approaches. Most urban arterials have an exclusive left-turn buffer at each approach to cater to the left-turning traffic. This prevents the left-turning traffic from blocking the through and right traffic until the buffer is filled.

Each approach is initially a single lane that fans out into a through-lane and an exclusive left-turn buffer. The left-turn buffers extend 60 meters and can hold 6-7 vehicles. There are two stop bar detectors per approach, one each for through/right and left turn lanes. There is one advance detector 90 meters from the intersection,

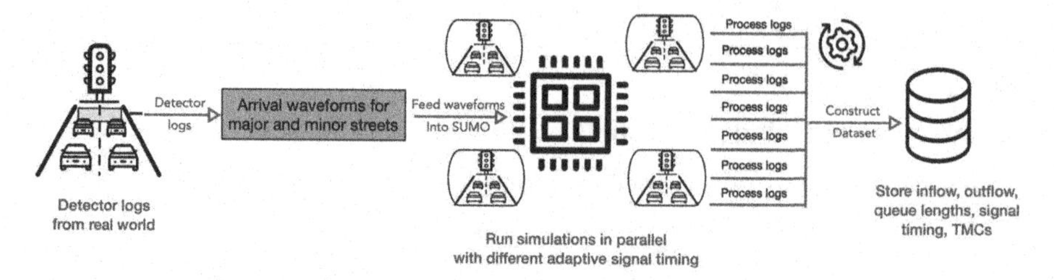

Figure 6.3: The arrival waveforms from real-world data from multiple intersections, are used to drive simulations. At any instant, several simulations will be running in parallel: each thread runs a simulation, processes the logs, and dumps the dataset into the file system.

just beyond the end of the left-turn buffer. This minimal configuration captures all the eight turn movements possible. Multiple lanes for each movement group can be handled by (a) aggregating detector counts per movement group and (b) training multiple models, one for each intersection geometry of interest. In this study, we only focus on the most general and minimal configuration.

In order to gather downstream data, we place gating traffic signals that mimic a downstream intersection 800 meters from the main intersection. They are simple one-phase signals without any side streets. Adding fully-fledged intersections with coherent real-world flows would have been computationally-expensive and will be included in a follow-up extension to this work. There are four such gating signals, one along each of the four outbound directions.

6.3.2 INPUT TRAFFIC GENERATION

Archived controller logs are used to construct inflow waveforms along all four directions. These waveforms are in turn used to inject vehicles into SUMO from four directions with an adaptive signal timing control. This way, we generate datasets whose inflow and outflow distributions are close to what we observe in the real world. For this, we used six months of controller logs from 90 intersections in the City of Orlando.

Thus, the main flow along the corridor-through and right directions will be between two to eight times the flow along the other streets. These ratios are based on the observed traffic flows in the recorded Orlando dataset.

We use advance detector logs from the Orlando dataset to generate vehicle flows at a 1-second resolution. We randomly sample flow patterns observed at these two detectors for the straight and side streets and ensure that they fall between the above-mentioned volume flow constraints. We program these arrival patterns into SUMO. These patterns are further shaped by the gating signals at the start of the four incoming approaches. This ensures variable platooning of volumes based on real-world data.

6.3.3 PARALLELIZATION

The data generation process makes use of the multiprocessing environment. At any instant, several simulations will be running in parallel: each thread runs a simulation, processes the logs, and dumps the dataset into the file system.

Each simulation generates logs that have information of every timestep of the simulation. These logs are processed, and the following information is stored:

- Waveforms at all the stop bar and advance detectors for all the approaches
- Waveforms at all the advance detectors of nearby intersections
- Signal timing information
- Turn movement counts for all the possible movements.

Within a simulation, after an initial simulation warm-up of 600 seconds, logs are extracted in windows of 1,000 seconds. These usually contain 8-9 complete cycles on average. Thus, each data exemplar consists of a set of waveforms of different signals and detectors, queue lengths, and turn movement counts for a window of 1,000 seconds (T = 200), aggregated at a 5-second resolution.

A large dataset of 5 million such exemplars is thus generated, accounting for 30 million traffic cycles of simulation. The dataset is then split for training and testing in the ratio of 70:30.

The creation of such a vast dataset involved considerable engineering effort. The entire pipeline was implemented in the Python programming language. A multiprocessing library was used to run up to 60 parallel instances of SUMO and preprocess output XML logs in batches. Numpy [54] and Dask [115] were then used to create vectors for training and testing. These vectors were stored in HDF5 format using the H5PY [26] library.

Implementation, training, and evaluation of the deep learning models were done using the PyTorch [110] library. University of Florida's HiPerGator supercomputing resources were used to train and test multiple models in parallel.

We use this dataset for modeling a universal approximator for predicting turn movement counts. The inputs to the model are:

- Detector actuations for all advance detectors (four detectors)
- Detector actuations at stop bar detectors for through-right and left buffers (eight detectors)
- Detector actuations at early detectors of nearby intersections (outflow detectors) (four detectors).

The information from these 16 detectors is used to predict turn movement counts. The detector actuations are aggregated to some level to construct detector waveforms. These detector waveforms are in turn used as input to the neural network models.

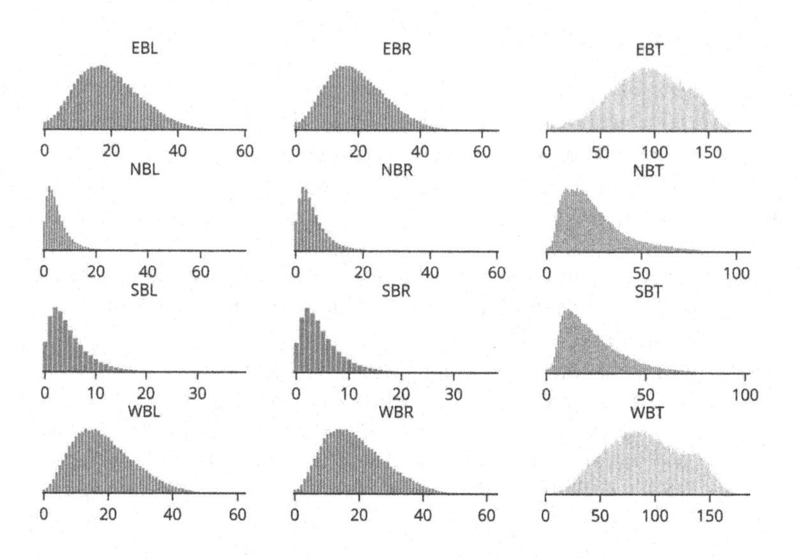

Figure 6.4: Turning movement count distributions of the dataset generated by simulation. Each subplot corresponds to a particular direction/movement. EBL: Eastbound left, EBT: Eastbound through, EBR: Eastbound right.

6.4 CONCLUSIONS

This chapter highlights the following:

1. Traffic Simulation Frameworks Importance: Traffic simulation frameworks play a crucial role in traffic engineering and machine learning applications. They enable the replication and analysis of complex traffic phenomena, formulation and evaluation of traffic management strategies, and provide dynamic environments for training machine learning models. These frameworks are indispensable tools for traffic engineering and machine learning tasks related to traffic management, autonomous driving, and more.
2. Types of Simulation Frameworks: Traffic simulation frameworks can be classified based on the complexity of modeling into macroscopic, microscopic, and mesoscopic simulators. Microscopic simulators, which simulate fine-grained vehicle behaviors, are particularly useful for analyzing intersections and corridors but are computationally-expensive.
3. Important Traffic Simulation Frameworks: Several traffic simulation frameworks were discussed, including VISSIM, SUMO, and CARLA. VISSIM offers a comprehensive platform for multimodal traffic flow simulation and supports various programming interfaces. SUMO, an open-source software, provides a lightweight and efficient solution for microscopic traffic simulation. CARLA, primarily built for autonomous driving research, offers realistic 3D rendering and can be co-simulated with SUMO or VISSIM.

4. Simulation Calibration: Calibration of simulation models is essential to ensure their accuracy and reliability. Techniques such as flow calibration and overspeeding calibration are used to adjust simulation parameters based on real-world observations. This process involves iterative refinement using data from loop detectors, probe trajectory data, and other sources to match simulated traffic flows with observed flows.
5. Dataset Generation: A detailed pipeline for generating traffic datasets using simulation frameworks was presented. This involved configuring intersections, generating realistic traffic flows, and parallelizing simulations to create large-scale datasets. The generated dataset was used for tasks such as predicting turning movement counts, which are vital for traffic management and optimization.

Overall, traffic simulation frameworks offer powerful capabilities for understanding and managing traffic dynamics, and their integration with machine learning techniques holds promise for addressing complex transportation challenges in urban environments.

7 Intersection Detector Diagnostics

Abstract: Current traffic signal controllers are capable of logging events (signal events, vehicle arrival, and departure) at very high resolutions (usually 10 Hz). The high resolution data rates enable the computation and study of various (granular) measures of effectiveness. However, without knowing the location of specific detectors on an intersection and the phases they are mapped to, a number of measures of effectiveness (of signal performance) cannot be evaluated. These mappings may not be available or up to date for many practical reasons (e.g., old infrastructure, mappings not machine readable, maintenance, addition of new lanes, etc.). In this chapter, we present an inference engine to map detectors to phases and distinguish between the stop bar and advance detectors or, in other words, infer the location of the vehicle detectors with reference to the intersection.

For computation of performance measures from the high-resolution controller logs, detector channel number to phase mappings are required. The phase represents the direction of traffic movement on an intersection. Additionally, we would like the mappings to help identify the location of a detector, such as advance or stop bar. In many practical situations, these mappings are missing (e.g., the infrastructure was built decades ago, and the mappings are not available in a machine-readable form) or incorrect (e.g., during maintenance or addition of new lanes, the contractor did not update the mappings). Additionally, there is no single, widely accepted standard for detector to phase mappings and it is possible that different agencies and vendors may have followed very different standards over the many decades of installing and maintaining the infrastructure. This problem and the need for standards in the detector to phase assignment process is outlined by C. M. Day, et all: in [29]

> *To measure performance of a signalized intersection in a meaningful manner or effect substantial performance improvements, it is critical to have vehicle detection sensors deployed. Since any detector event must be understood as a vehicle arriving at some right-of-way and distance at the intersection, and any phase event must be understood as an interval of time elapsed for some movement, the detector mapping process is the critical step in guiding how signal output states and vehicle arrivals are to be interpreted relative to each other.*

Our goal in this chapter is to show how one can derive, in a data driven manner, the mapping of detectors to phases and to classify detectors as stop bar detectors or advance detectors based only on events in the high-resolution controller logs. Broadly, these logs include the following events: changes in the signaling state (e.g., green, yellow, or red for vehicles, and walk, flashing do not walk, and do not walk for pedestrians) and changes in the detector state (based on whether the detection area is occupied or not).

Our algorithms are based on the following observations:

- During normal traffic conditions, the traffic crossing a detector when the corresponding phase is active (or green) will be higher than the traffic crossing detectors corresponding to other phases.
- During times with very low traffic volumes, the sequence of timestamps of reported vehicle arrival and departure events can be used to separate advance detectors from stop bar detectors.

A completely automated, data driven, machine learning methods that derive these mappings using several months of controller event logs is presented in the following sections. This consists of following key steps:

- Automatically decompose the SPaT data into timing cycles and signal timing patterns, and then cluster these cycles based on similar phase timing patterns. This is necessary because combining data across multiple, distinct types of patterns can lead to higher discrimination between green versus red behavior. For this purpose, we developed frequent item-set mining based on n-grams (i.e. a collection of ‘n’ number of successive text tokens) and then applied clustering algorithms to derive clusters of similar cycles. For medium volume cycles, we developed algorithms that derive features that correlate the behavior of traffic at a given detector with the phase activity corresponding to the signal timing. Effectively, these features use green versus red departure information to derive the potential feasible phases for each detector.
- Using the features described in the previous step, we develop two algorithms. The first one uses frequent set mining within a given cycle cluster to derive a consistent mapping for that cluster and then combine detector mappings derived for each cluster of similar cycles to arrive at the overall mapping and the detector type. The second one uses a neural network approach to assign a detector to a particular phase with a classification setting. We cross-check the results obtained by the two models for consistency.
- For really low volume cycles (e.g., during night-time), when only one vehicle is potentially crossing the intersection, the time of arrival and departure along with signal timing data can be used to derive ordering of detectors. Using this information, we develop approaches for differentiating whether a detector is a stop bar or advance detector.

We first define the problem formally. This is followed by a detailed description of each of the above steps.

7.1 PROBLEM DEFINITION

Inductive detectors and signal controllers installed at intersections, as shown in Figure 7.1, collect high resolution data, which provides information about the vehicle arrival and departure events and the corresponding signal timing state during these events. These detectors are placed on each lane and are of two types: advance and stop bar detectors. Most of the intersections have four approaches and eight phases or directions of vehicular movement. A signal timing cycle is defined as:

> *the total time to complete one sequence of signalization for all movements at an intersection. In an actuated controller unit, the cycle is a complete sequence of all signal indications.* [134]

Figure 7.1 is a representation of the kind of intersection we use for demonstrating our methodology throughout the rest of the chapter. This intersection is referred to as intersection *I*. The problem of mapping of detectors to phases can be represented in the form of a matrix. Here, the rows represent the phases and the detectors are

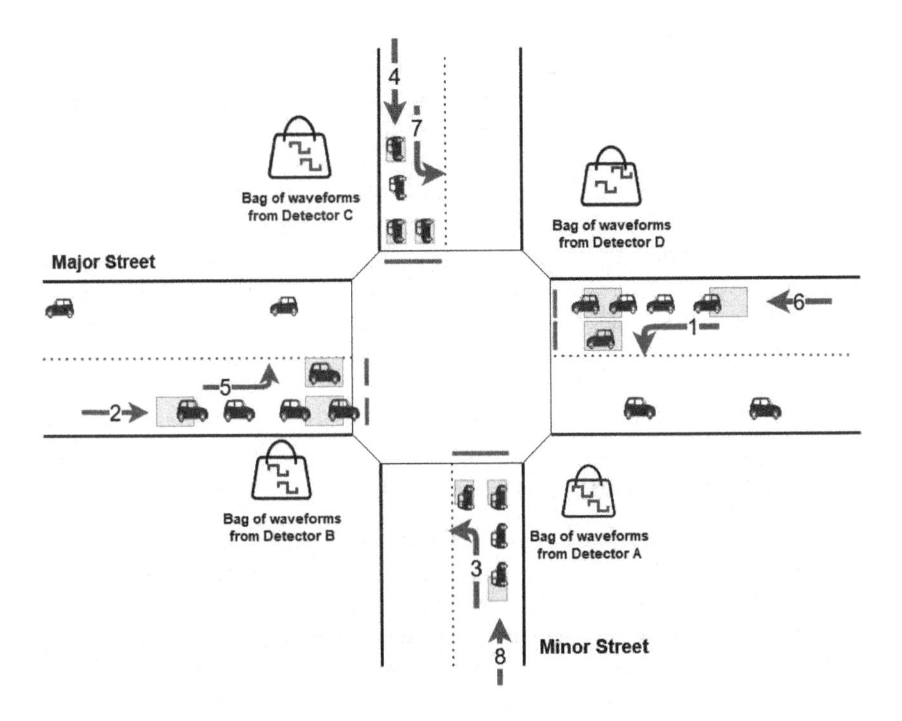

Figure 7.1: An example illustrating intersection I, with the 8 vehicular phase movements, the signal timing state, and the data being generated by the detectors on each lane.

represented as columns. If a detector does not belong to a particular phase, the corresponding matrix element is 0. However, a matrix entry of 1 represents that the detector may belong to the corresponding phase.

In the absence of any information, all the elements in this assignment matrix are 1, i.e. a situation where a detector may belong to any phase. our goal is to sparsify this assignment matrix using our algorithms. Table 7.6 shows the final state of the matrix using one of our algorithms. Each phase may have multiple detectors; hence, there would be many 1s in each row. However, usually, there should be only a few 1s or a single 1 per column because a detector often belongs to only one phase. If non-conflicting phases always start and end together, then it is much more difficult to assign a detector located on one of these non-conflicting phases to a unique phase based on the controller logs alone.

For an intersection, for each detector, we sample waveforms (i.e. binary time series of detector actuations with '0' representing 'OFF' and '1' representing 'ON' states) from different time periods of the day and for several days. So each detector is represented by a bag, and the bag has several thousands of waveforms. Our task is to be able to classify each bag to one of the eight phases. If we look at individual detector waveforms, they may look similar for different detectors but the characteristics of the bag as a whole are being used for assigning it to a particular phase.

The idea of using multiple, bagged instances to classify the bag is similar to the Multiple Instance Learning (MIL) problem. MIL is a type of weakly supervised learning where training instances are arranged in bags (or sets). In this setup, a label is given to the entire bag instead of individual exemplars.

But in many applications the standard assumptions are too restrictive and these assumptions can be relaxed. We use the fact that the inherent data distribution is different for each of the bags. Also, each bag contains a sufficient number of instances that are representative of a particular class (in our problem statement, a phase is represented by a class) and that class is more predominant than other classes. This setup is a natural fit for our problem, as by looking at individual detector waveforms alone we can't find any distinctive feature. However, by considering the whole bag of waveforms we can assign the bag to a particular phase.

7.2 FEATURE EXTRACTION

We consider cycles from time periods with moderate traffic activity. This is done by filtering out periods of very low (e.g., nighttime) and very high (e.g., rush hour) traffic. The rationale for doing this is that if there is little traffic, the detectors would not be activated, and on the other hand, if there are many vehicles at the intersection, then most detectors will be activated too frequently, which may make the differences in activation counts less perceptible. The fundamental observation that forms the bases of both our algorithms is that a detector will record more departures when the corresponding phase is green (or yellow) when compared to the number of recorded departures when the same phase is red. Thus, the expectation is that within a timing cycle, for each detector, if we were to assign a positive vote to each reported vehicle departure when phase 6 is green and negative vote to each vehicle departure event

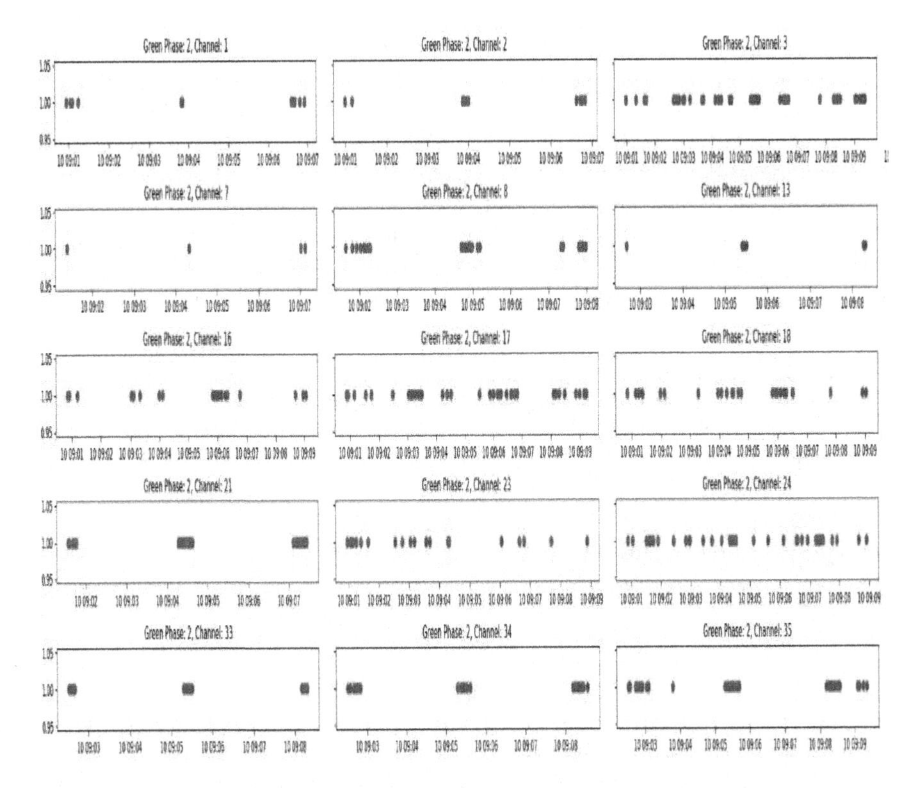

Figure 7.2: A visualization of departures reported by some detectors of intersection I for multiple cycles. Certain detectors report departures only when phase 2 is green. This is a strong signal that these detectors belong to Phase 2.

when phase 6 is red, all the detectors which may belong to phase 6 will be left with a positive vote count by the end of the timing cycle.

7.2.1 SIGNAL TIMING PATTERN IDENTIFICATION AND CLUSTERING

The first step in this process is to automatically decompose the event logs from a single intersection into SPaT (Signal Phase and Timing) events and detection events. Next, we need to identify the signal timing patterns deployed at the intersection and further decompose this into signal timing cycles. This is needed because by combining observations and conclusions drawn from different signal timing patterns leads to much higher discrimination between green vs red behaviour for a particular detector. To this end, we perform frequent item-set mining on the n-grams and use the results for clustering. Frequent item-set mining on n-grams will inform us the frequency that each n-gram sequence appears in the entire set of phase events. This is illustrated with an example below. A representation of the signal phase and timing data is created for this purpose by only noting the phase or phases that were reported to be green simultaneously. For example, if phases 2 and 6 receive green time simultaneously, followed by the phases 1 and 6; 3 and 8; 4 and 8; and 4 and 7, this is represented as a sequence of changes: '2,6', '1,6', '3,8', '4,8', '4,7', and so on. This simplified representation of the SPaT data can then used to identify signal timing cycles by application of n-grams.

n-grams can be defined as a continuous sequence of n words (or elements) in a text (or a collection of elements) and have been extensively used in the fields of computational linguistics and probability. The representation of the SPaT data described above can used to find n-grams and hence automatically deduce the signal timing patterns deployed at an intersection. For example, a frequently repeating sequence of the 6-gram: ('2,6', '2,5', '3,7', '4,7', '4,8', '1,6') indicates that is is one of the dominant signal timing pattern used at the intersection. A few examples of n-grams found in SPaT data are shown in Figure 7.3. Whereas, Table 7.1 presents the most frequent 6-grams, or timing patterns, for the same intersection.

'2,5', '2,6', '1,6', '4,7', '4,8', '3,8', '2,6', '2,5', '3,7', '4,7', '4,8', '1,6', '2,5', '2,6', '1,6', '4,7', '4,8', '3,8', '2,6', '2,5', '3,7', '4,7', '4,8', '1,6', '2,6', '1,6', '3,8', '4,8', '4,7', '2,5', '2,6', '2,5', '3,7', '4,7', '4,8', '1,6', '2,6', '1,6', '3,8'

Examples of 2-grams: '2,5', '2,6' | '1,6', '4,7' | '2,6', '2,5' |

'2,5', '2,6', '1,6', '4,7', '4,8', '3,8', '2,6', '2,5', '3,7', '4,7', '4,8', '1,6', '2,5', '2,6', '1,6', '4,7', '4,8', '3,8', '2,6', '2,5', '3,7', '4,7', '4,8', '1,6', '2,6', '1,6', '3,8', '4,8', '4,7', '2,5', '2,6', '2,5', '3,7', '4,7', '4,8', '1,6', '2,6', '1,6', '3,8'

Examples of 3-grams: '2,5', '2,6', '1,6' | '1,6', '4,7', '4,8' | '2,6', '2,5', '3,7' |

'2,5', '2,6', '1,6', '4,7', '4,8', '3,8', '2,6', '2,5', '3,7', '4,7', '4,8', '1,6', '2,5', '2,6', '1,6', '4,7', '4,8', '3,8', '2,6', '2,5', '3,7', '4,7', '4,8', '1,6', '2,6', '1,6', '3,8', '4,8', '4,7', '2,5', '2,6', '2,5', '3,7', '4,7', '4,8', '1,6', '2,6', '1,6', '3,8'

Examples of 6-grams: '2,5', '2,6', '1,6', '4,7', '4,8', '3,8', | '2,6', '2,5', '3,7', '4,7', '4,8', '1,6',

Figure 7.3: 2-,3-, and 6-grams in the SPAT data. We use this approach to infer signal timing plans from SPAT messages.

Table 7.1

Most frequent 6-grams found in the SPAT messages from a controller. These correspond to the predominant signal timing plans deployed at this controller.

6-Gram	Frequency	Timing Pattern
('2,5', '2,6', '1,6', '4,7', '4,8', '3,8')	117	Pattern 1
('2,6', '2,5', '3,7', '4,7', '4,8', '1,6')	84	Pattern 2
('2,6', '1,6', '3,8', '4,8', '4,7', '2,5')	67	Pattern 3
('2,5', '2,6', '1,6', '4,7', '4,8', '2,5')	17	Pattern 4

7.2.2 FEATURE DETECTION

The vehicle actuation on a detector can be considered as a pulse waveform and can be represented as a 1D vector, with T components. Here T refers to the length of time a particular detector's data are being considered, with each component being aggregated for a 5-second period. In our work, T = 150, i.e., each vector corresponds to 750 seconds of data (roughly 6-7 cycles), aggregated for 5-second periods. The time gap between two vehicles (headway) near an intersection is usually 2 seconds, which leads to an average of 2-3 vehicles per 5-second period. We find this level of aggregation sufficiently expressive to capture platoon dynamics.

Signal timing information can be represented as an eight-dimensional vector (where the 8 dimensions correspond to 8 phases), each with T components. A feature vector of T components for each phase is created to represent a cycle where each time interval has a value of +1 when the phase is active and -1 when the phase is inactive. To exploit the fact that signal timing is correlated with detector actuations, we construct the following features using the eight signal timing vectors and each detector waveform. We take the dot product of each detector waveform with each of the eight signal timing vectors and compute dot product scores of the detector waveform with each phase. The intuition is that the dot product score will be high for the actual phase that the detector belongs to.

If the events were aggregated to a resolution of 5 seconds and the timing cycle being analyzed was of the length of 30 seconds and the feature vector for phase 6 with six entries is as follows: [-1,-1,+1,+1,+1,-1] then this representation would indicate that phase 6 was green (or active) between seconds 11 and 25 and was red (or inactive) for the rest of the cycle. More formally, a detector waveform is represented as a 1D vector,

$$d_i = (c_1, c_2, \ldots, c_{T-1}, c_T),$$

where,

c_t is vehicle count for time interval t
T is number of time intervals

Table 7.2

Partial assignment matrix based on Pattern 1. This assignment matrix is derived from Pattern 1 in Table 7.1. Detectors (columns) are assigned tentative phases based on the number of departures on green vs. red, but it is difficult to distinguish between phases that were green together, e.g., phases 1 and 6.

Detectors	1	2	7	8	13	14	15	16	17	18	21	26	33	34	40	41	45	46
Phase 1	464	578	-86	-790	-114	-218	312	307	189	88	-638	-589	-378	-466	-963	-849	254	189
Phase 2	-464	-578	-86	-790	114	218	-312	-307	-189	-88	-638	-589	378	466	-963	-849	-254	-189
Phase 3	-464	-578	86	-790	-114	-218	-312	-307	-189	-88	492	439	-378	-466	-963	-849	-254	-189
Phase 4	-464	-578	-86	790	-114	-218	-312	-307	-189	-88	-492	-439	-378	-466	963	849	-254	-189
Phase 5	-464	-578	-86	-790	114	218	-312	-307	-189	-88	-638	-589	378	466	-963	-849	-254	-189
Phase 6	464	578	-86	-790	-114	-218	312	307	189	88	-638	-589	-378	-466	-963	-849	254	189
Phase 7	-464	-578	-86	544	-114	-218	-312	-307	-189	-88	-638	-589	-378	-466	697	729	-254	-189
Phase 8	-464	-578	86	-544	-114	-218	-312	-307	-189	-88	638	589	-378	-466	-697	-729	-254	-189

The signal timing vector for waveform i, phase j, $\vec{st}_i^j$, is represented as:

$$\vec{st}_i^j = [1, -1, -1, \dots, 1, 1],$$

1 at index t indicating phase j is active for that time interval.

Then the feature vector for waveform i is the dot product of d_i with signal timing vector $\vec{st}_i^j$

$$\vec{F}_i = [\vec{d}_i \cdot \vec{st}_i^1, \vec{d}_i \cdot \vec{st}_i^2, \dots, \vec{d}_i \cdot \vec{st}_i^8]$$

7.3 DATA MINING APPROACH

We first present a data mining approach for the problem of detector-to-phase mappings (Figure 7.4). We consider cycles from time periods with moderate traffic activity. This is done by filtering out periods of very low (e.g., night time) and very high (e.g., rush hour) traffic. The rationale for doing this is that if there is little traffic, the detectors would not be activated and, on the other hand, if there are many vehicles at the intersection, then most detectors will be activated too frequently, which may make the differences in activation counts less perceptible.

We also look for different timing patterns on the intersection such that during the different patterns, pairs of non-conflicting phases are served separately. This makes it possible to disambiguate the detector mappings further by considering the number of departure votes over these differing signal timing patterns. A visualization of the number of departures for multiple cycles is presented in Figure 7.2 where certain detectors report departures only when phase 2 is green. This can be seen as an indication that those detectors may be assigned to phase 2. Additional details and observations are presented in steps below.

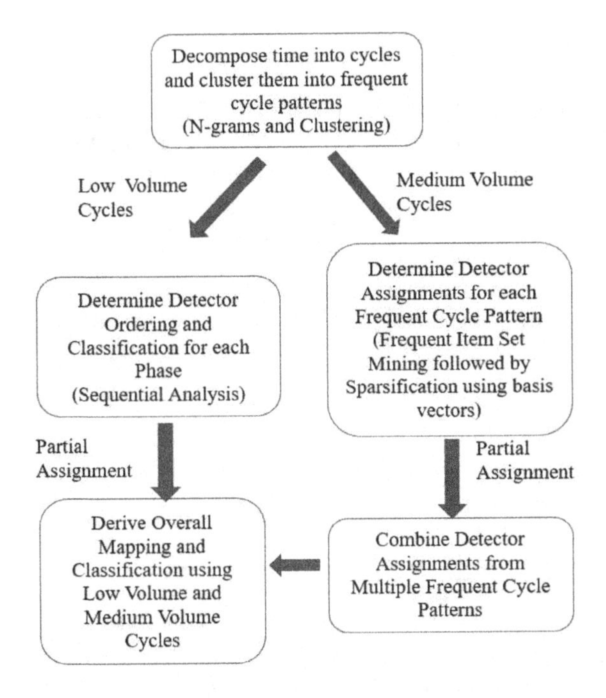

Figure 7.4: Flow diagram for our white box inference engine. We separate out the low volume and medium volume data from the logs. The medium volume data is used for partial assignments, and low volume data is used to distinguish between stop bar and advance detectors and to enable ordering of detectors.

7.3.1 INFERENCE FROM A SINGLE CLUSTER

We start with the premise that inferences drawn from one timing cycle will not be error free. This is explained by the following observation: The demand on a particular phase or direction may be abnormally high during the cycle. Combining the results obtained from multiple timing cycles and getting consistent results increases the confidence in the results while reducing errors. To this end, we cluster or group the timing cycles with the same ordering of phases (same timing pattern) and derive results from each cluster or timing pattern separately. This process is explained using the most popular patterns, Pattern 1 and Pattern 2, from Table 7.1. For two cycles that have the same pattern, we take a set union of the results for these two cycles. The rationale for taking a union is that the assignments may be slightly different for the two cycles based on specific traffic conditions during the occurrence of these cycles. For example, in one case, the volume of traffic may be low, resulting in the mapping of fewer detectors. Figures 7.2 and 7.3 show the detector-to-phase mappings obtained from Pattern 1 and Pattern 2, respectively, by applying the union operation on all

Table 7.3

Partial assignment matrix based on Pattern 2. This assignment matrix is derived from Pattern 2 in Table 7.1. We combine the results from Pattern 1 (Table 7.2) and Pattern 2 in Figure 7.5 to get better mappings.

Detectors	1	2	7	8	13	14	15	16	17	18	21	26	33	34	40	41	45	46
Phase 1	446	551	-152	-923	-187	-250	449	385	187	82	-819	-718	-432	-467	-1062	-1040	614	518
Phase 2	-446	-551	-152	-923	187	250	-449	-385	-187	-82	-819	-718	432	467	-1062	-1040	-614	-518
Phase 3	-446	-551	152	-923	-187	-250	-449	-385	-187	-82	-819	-718	-432	-467	-1048	-1040	-614	-518
Phase 4	-446	-551	-152	923	-187	-250	-449	-385	-187	-82	819	718	-432	-467	1048	1040	-614	-518
Phase 5	-446	-551	-152	-923	187	250	-449	-385	-187	-82	-819	-718	432	467	-1062	-1040	-614	-518
Phase 6	446	551	-152	-923	-187	-250	449	385	187	82	-819	-718	-432	-467	-1062	-1040	614	518
Phase 7	-446	-551	152	-291	-187	-250	-449	-385	-187	-82	-819	-718	-432	-467	-116	-168	-614	-518
Phase 8	-446	-551	-152	291	-187	-250	-449	-385	-187	-82	819	718	-432	-467	116	168	-614	-518

cycles with the same pattern. Tables 7.4 and 7.5 show these mappings for the Pattern 1 and Pattern 2, respectively.

7.3.2 INFERENCE FROM MULTIPLE CLUSTERS

For two or more clusters consisting of cycles with different patterns, we use association-rule mining metrics to combine results. Specifically, for each detector (Det) that needs to be assigned a phase (p), we compute the support for each phase across all the clusters. In association-rule mining [1] and when processing transactional data, a rule is defined as an implication statement $X \implies Y$ derived from a set of transactions T. Support is a measure of frequency (or relative importance) of the set X and is usually defined as the proportion of transactions in T that include the set X. In our case, support can be defined as the fraction of times a detector was assigned to a particular phase across all signal timing patterns. An illustration of the use of multiple clusters to obtain disambiguated results can be derived by using the results from Tables 7.4 and 7.5 together. Here, detector 7 is assigned to Phases 3 and 8 in Figure 7.2 because both Phase 3 and Phase 8 have the same score. Similarly, detector

Table 7.4

Partial Mapping deduced using only *Pattern* 1. This is a transactional representation of the results presented in Figure 7.2.

Detector Number	Possible Phases
13,14	Phase 2 or Phase 5
1,2	Phase 6 or Phase 1
21,26	Phase 8
8,40,41	Phase 4 or Phase 7
7	Phase 3 or Phase 8

Table 7.5
Partial Mapping deduced using only *Pattern* 2. This is a transactional representation of the results presented in Figure 7.3.

Detector Number	Possible Phases
13,14	Phase 2 or Phase 5
1,2	Phase 6 or Phase 1
7	Phase 3 or Phase 7
21,26	Phase 4 or Phase 8

7 is assigned to Phases 3 and 7 in Figure 7.3. By computing the support (as defined above), since Phase 3 was present in both transactions, we can assign detector 7 to Phase 3. Following the same process, it is easy to assign detectors 21 and 26 to Phase 8 and detector 1 to Phase 1.

Thus, applying a combination of the set union (within a cluster) and association-rule mining techniques (across multiple clusters: Figure 7.5), we arrive at the final assignment results shown in Figure 7.6.

7.3.3 CLASSIFYING DETECTORS

To infer the relative locations of the detectors from the stop bar, we turn our focus to very low volume-timing cycles. The key intuition here is that detectors are always triggered in the reverse order of their distance from the stop bar by a single car. To this end, we select the timing cycles with exactly n (n =3) recorded departure events within a short duration of time. In most cases, a single car can cross no more than 3 detectors while crossing the intersection. Hence, n is chosen to be 3. Next, we arrange these recorded departures as ordered item sets and compute the most frequent item sets. For example, detector 17 records a departure event, which is then followed by departure event recorded on detector 15 and then on detector 45. Assume that this particular order of events (or ordered set) is observed in the data with a high frequency. This observation can then be used to deduce the relative location of these detectors from the stop bar. This technique was found to be most useful on the major

Pattern 1	Pattern 2	Support	Result
Det 7 → P3,P8	Det 7 → P3, P7	P3 = 1 P8 = .5, P7 = .5	Det 7 → P3
Det 21,26 → P3,P8	Det 21,26 → P4, P8	P8 = 1 P3 = .5, P4 = .5	Det 21,26 → P8

Figure 7.5: We use well-known association-rule mining techniques to combine the partial mappings derived using data from different timing patterns.

Channels	1	2	3	4	6	7	8	13	14	15	16	17	18	19	20	21	26	33	34	35	36	38	39	40	41	42	43	44	45	46	49	50	51	64
Phase 1	1	1	0	0	0	0	0	0	0	0	0	0	0	0	0	0	0	0	0	0	0	0	0	0	0	0	1	1	0	0	0	0	0	0
Phase 2	0	0	1	1	0	0	0	1	1	0	0	0	0	0	0	0	0	1	1	1	1	0	0	0	0	0	0	0	0	0	0	0	0	0
Phase 3	0	0	0	0	0	0	0	0	0	0	0	0	0	0	0	0	0	0	0	0	0	0	0	0	0	0	0	0	0	0	0	0	0	0
Phase 4	0	0	0	0	0	0	1	0	0	0	0	0	0	0	0	0	0	0	0	0	0	0	0	1	1	1	0	0	0	0	0	0	0	1
Phase 5	0	0	0	0	0	0	0	1	1	0	0	0	0	0	0	0	0	0	1	1	0	0	0	0	0	0	0	0	0	0	0	0	0	0
Phase 6	1	1	0	0	0	0	0	0	0	1	1	1	1	1	1	1	1	0	0	0	0	1	1	0	0	0	1	1	1	1	1	1	0	0
Phase 7	0	0	0	0	0	0	0	0	0	0	0	0	0	0	0	0	0	0	0	0	0	0	0	0	0	0	0	0	0	0	0	0	0	0
Phase 8	0	0	0	0	0	0	0	0	0	0	0	0	0	0	0	0	0	0	0	0	0	0	0	0	0	0	0	0	0	0	0	0	0	0

Figure 7.6: Final mapping or assignment matrix deduced using a combination of *Pattern* 1 and *Pattern* 2. We start with the assumption that every detector may be assigned to any phase. Hence, we assume that all the cells in the assignment matrix are green. Techniques presented in this chapter can be used to sparsify the initial matrix and arrive at this final mapping where green cells represent likely assignment.

phases (i.e. Phases 2 and 6). The results for the ordering of detectors derived using this technique for one intersection are presented in Table 5.4.

7.4 NEURAL NETWORK APPROACH

Neural network approaches have been successfully used for classification. Generally, these methods learn input-output relationships using a feed-forward network. For multi-class classification approaches, the number of outputs is equal to the number of classes.

Our goal in this section is to classify a bag of waveforms assigned to a given detector for two purposes: phase assignment and advance vs. stop bar classification. We train two different neural networks for this purpose. Each waveform in this bag corresponds to departure information at the detector for one cycle and corresponding phasing information for that cycle. This can be converted into an input-output pair

Table 7.6

Frequent item sets observed using very low volume observation periods. These results are in agreement with the results presented in Figure 7.6. They also help us distinguish between stop bar and advance detectors.

Item Set	Frequency	Green Phases
6,4,36	35	Phase 2 and 6
17,15,45	26	Phase 2 and 6
18,16,46	23	Phase 2 and 6

Phases →

Timing Plan ↓

	4.0	8.0	2.0	6.0	7.0	1.0	3.0	5.0
0.0	{5}	{5}						
1.0	{6, 5, 10, 11, 12}	{11, 12}	{2, 7, 14, 15, 3, 16, 13}	{2, 7, 14, 15, 3, 16, 13}	{6, 10}			
2.0	{11, 5, 16}	{11, 5, 16}	{2, 7, 14, 15, 3, 13}	{2, 7, 14, 15, 3, 13}				
3.0	{6, 11, 5, 12}	{6, 11, 5, 12}	{2, 14, 15, 3, 16, 13}	{2, 14, 15, 3, 16, 13}	{4, 10}	{7, 1}	{4, 10}	{7, 1}
4.0	{6, 5}	{6, 4, 5, 16, 11, 12}	{14, 15, 3}	{14, 15}			{11, 4, 12, 16}	
5.0	{6, 5, 12}	{6, 5, 12}	{2, 14, 15, 3, 16, 13}	{2, 14, 15, 16, 13}				
6.0	{6, 5, 10, 11, 12}	{11, 12}	{2, 7, 14, 15, 16, 13}	{2, 7, 14, 15, 3, 16, 13}	{6, 5, 10}	{1}		
7.0	{11, 5, 14}	{11, 5, 14}	{16, 15, 3}	{16, 15, 3}		{1}		{1}
8.0	{6, 12}	{12}	{2, 7, 14, 15, 3, 16, 13}	{7, 14, 15, 16, 13}	{6}			
9.0	{11, 5, 12}	{11, 12}	{2, 7, 14, 15, 3, 16}	{2, 14, 15, 16}	{5, 10}		{10}	
10.0	{6, 5}	{6, 4, 5, 11, 12}	{2, 7, 14, 15, 3, 16, 13}	{2, 7, 14, 15, 3, 16, 13}		{1}	{4, 12}	{1}
11.0	{6, 10}	{6, 4, 12}	{2, 7, 14, 15, 3, 16}	{15, 16}	{10}		{4, 12}	{7, 2, 14}
12.0	{6, 5, 10}	{11, 4, 12}	{7, 2, 13, 14}	{1, 16, 15, 3}	{10}	{1}	{4}	
13.0	{6, 5, 12}	{6, 11, 5, 12}	{7, 2, 13}	{1, 16, 15, 3}		{1}		{7, 2, 13}
14.0	{6, 5, 10, 11, 12}	{11, 12}	{7, 2, 13, 14}	{1, 16, 15, 3}	{10}	{1}		
15.0	{6, 5, 14, 11, 12}	{6, 5, 14, 11, 12}	{7, 2, 13, 16}	{1, 16, 15, 3}		{1}		{7}

Detector assignments from different signal timing plans

Phase	1	2	3	4	5	6	7	8
Detector	1	2,3	4	5,6	7	8,9	10	11,12

Previously Known

Phase	1	2	3	4	5	6	7	8
Detector	1	2,3, 13,14	4	5,6	7	15,16	10	11,12

Predicted

Figure 7.7: An example of an intersection in City 1, where our techniques were used to derive the correct mappings. We found that all the existing mappings were correct. However, mappings for two of the detectors (13 and 14) were missing.

where the input corresponds to the feature matrix (described in Section 7.2) and the output corresponds to the actual phase the detector belongs to.

Rather than directly developing a neural network approach for bag classification, we used a two-phase approach. In the first phase, we develop a classifier that predicts which bag a waveform belongs to. This classifier has a high level of inaccuracy because a particular waveform may belong to multiple bags (effectively arrival-departure patterns are similar for two detectors for the same phasing information). However, because there are enough waveforms that are somewhat unique or are more repetitive for a particular detector, this classifier can then be used to score an entire bag of waveforms corresponding to a detector and make accurate assignments. Thus, even if this classifier correctly predicts the correct phase at an accuracy of $(1 + k)$ times the accuracy of assigning randomly to one of the phases, given a large enough bag, it can correctly assign the bag to the right phase.

For each waveform in a bag, we calculate a feature vector as described in Section 7.2. These are used as input to the classifier for that waveform. The classifier for each waveform is labeled as the phase to which the detector belongs. Thus, the set of

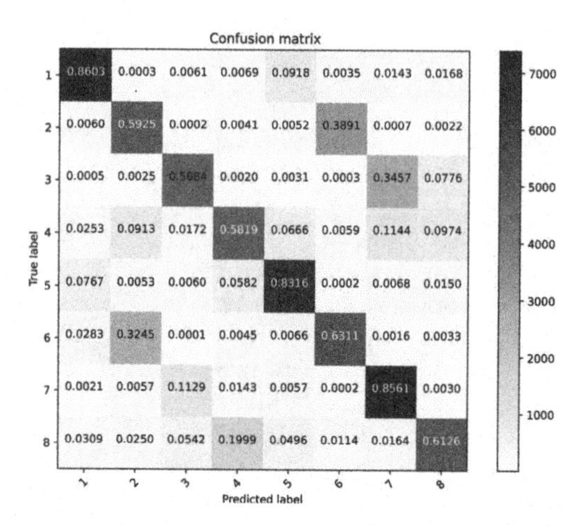

Figure 7.8: Plot showing confusion matrix for the neural network. We can see that the predicted labels match the ground truth for most samples. We can also see that phases that are typically synchronised are more challenging to distinguish. The key observation is that if the size of the bag is more than a few hundred cycles, the correct phase will have a significantly larger number of votes, and it is easier to determine the correct phase for every detector.

labels for $\{\vec{d_i}\}$ is $[j, j, \ldots, j]_N$, where j is the phase it belongs to. The input-output combination for each waveform from the set is assigned to the corresponding target class, and the neural network is trained to predict for each waveform.

During the testing phase, let $\{\vec{d_i}\}$ be the set of detector waveforms and $\{o_i^N\}$ be their corresponding outputs. The o_i with the highest number of votes is taken as the correct phase assignment for the detector. As discussed earlier, the neural network may make many mistakes for the individual waveforms in the entire set, but even an accuracy of 30% should be good enough to be able to find the correct phase (a random assignment would correspond to 12.5% accuracy). We use a fully connected network with three hidden layers with 25, 12, and 15 hidden units, respectively. The training converged in less than 200 iterations; and the testing error was only slightly more than the training error. The confusion matrix for the final results is presented in Figure 7.8. A similar neural network approach is used to determine advanced vs. stop bar prediction. The results are presented in Figure 7.9.

7.5 VERIFICATION

The approaches described above can be used to determine a sparse assignment matrix of detector-to-phase mapping. If this predicted mapping conflicts with the previously known mapping, then the derived mapping can be effectively used to provide suitable corrections. In this section, we describe the above two approaches as applied to data from two cities.

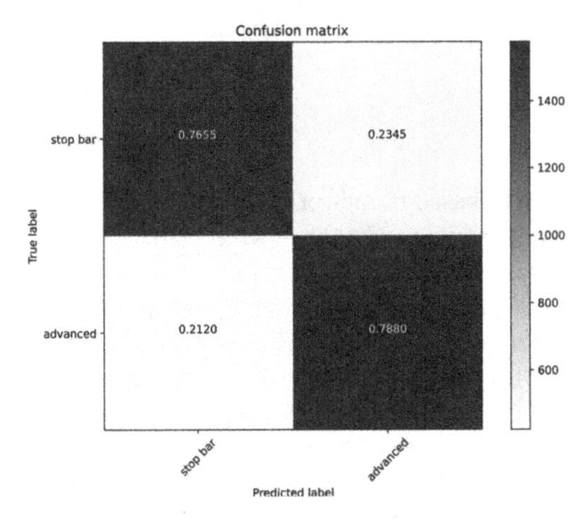

Figure 7.9: Plot showing confusion matrix for stop bar vs. advance detector classification. The key observation is that if the size of the bag is more than a few hundred cycles, we should be easily able to distinguish between whether a detector is stop bar or advance. For example, for a stop bar detector with 100 waveforms, the expected number of votes will be 76 for stop bar and 24 for advance.

7.5.1 CASE STUDY 1

We had detector phase mapping information available for 27 intersections. First, the data mining approach was used to determine the detector mappings and manually compared against the mappings available. The mappings that we found were mostly consistent with the provided mappings. However, we found detectors that were no longer active and detectors where the mappings were out of date. These findings were confirmed by the city traffic engineers.

The neural network approach needs labeled data for training. Using labeled data from seven detectors derived from the direct mapping (DM) approach, we trained the neural network to predict the correct phase (Figure 7.7). We then used this model to predict the mappings for the remaining 20 intersections. These were consistent with the mappings found using the DM approach.

7.5.2 CASE STUDY 2

Next, we demonstrated the ability of the models to generalize to intersections in another city. We used both approaches to predict the detector-to-phase assignments for 20 intersections in another city. The neural network was able to correctly predict the mappings without any retraining (i.e., we used the model built from City 1). The predictions from the two models were consistent with each other. We were able to identify several errors in the existing mappings.

To summarize, the key problems with sensor data, data quality and cleaning issues were explored in this chapter. We also propose a partial solution for some of these problems. In the next two chapters, we detail the key components of our decision support system.

7.6 CONCLUSIONS

We have described a MIL based framework to derive detector to phase mappings and used it to verify partial mappings when they were available. We also presented a way to distinguish between stop bar and advance detectors.

First, we presented a data mining approach that uses low volume cycles (e.g., during night-time when only one vehicle is present) to derive whether a detector is advance or stop bar. It uses the time of arrival and departure along with signal timing data to derive the ordering of detectors. We also presented algorithms that can use green versus red departure information to derive a small subset of phases for each detector. We then use association-rule mining to combine information from multiple cycles and assign a detector to a phase.

Next, we also descrived a neural network-based approach that exploits the relationship between signal timing and vehicle departures to train a neural network to be able to assign a detector to a particular phase with a classification setting. Furthermore, we extended this approach to determine whether a detector is advance or stop bar.

8 Intersection Performance

Abstract - Traffic signals are installed at road intersections to control the flow of traffic. An optimally operating traffic signal improves the efficiency of traffic flow while maintaining safety. The effectiveness of traffic signals has a significant impact on travel time for vehicular traffic. There are several measures of effectiveness (MoE) for traffic signals. In this chapter, we present a work-flow to automatically score and rank the intersections in a region based on their performance, and group the intersections that show similar behavior, thereby highlighting patterns of similarity. In the process, we also detect potential bottlenecks in the region of interest.

The primary function of the traffic controllers is to ensure safety by eliminating (or managing) conflicts and maximizing the flow of traffic while adhering to the safety constraints. The newest generation of traffic signal controllers, which are ATC[1] compliant, can record both signal phasing information and vehicle arrivals and departures at a very high resolution (10 Hz). As discussed above, this has resulted in the development of many new measures of effectiveness (MoEs) including the development of the automated traffic signal performance measures (ATSPM)[2] and other related efforts. These tools—that enable traffic engineers to quantitatively study the performance of signalized intersections—are being developed and deployed in cities and regions all across the United States. This newer generation of controllers, combined with the ATSPM systems, provide vastly improved monitoring capabilities as compared to the older generation systems, which were based on coarse grained data (typically at 15-minute intervals). The previous chapter highlighted the need for detector to phase mapping which is critical to draw any conclusions based on the the controller logs.

In this chapter, we describe a framework that combines processing of high resolution controller log data pertaining to certain performance measures with modern data mining and machine learning techniques to produce the following outcomes:

1. Automatically learn a compact representation of signal performance and behavior both in terms of the signal's capacity to serve demand and coordination of signal with upstream and downstream intersections.
2. The compact representation enables the grouping of signals based on similar demand patterns and performance over time and space as well as the ordering of these groups in terms of observed performance.
3. Our models can be used to discover significant changes in signal performance and hence to estimate the need to update signal timing plans.
   In other words, detect temporal changes in signal performance over multiple weeks or months and detect periods with changes in many signals.

[1]https://www.ite.org/technical-resources/standards/atc-controller/

[2]https://udottraffic.utah.gov/atspm/

4. Automatic generation of human understandable descriptions for each of these performance groups or clusters and ways of further compacting the performance measure for each of these groups.

8.1 UNSUPERVISED DATA SUMMARIZATION

vector space representation and clustering for the summarization of signalized intersection data. We first describe the Measures of effectiveness (MoE) pertinent to the problem. After setting up the MoEs in this space, we describe a denoising and compression approach which utilizes the SVD and higher order SVD. This is followed by a description of the clustering approaches used.

8.1.1 PERFORMANCE MEASURES

In order to quantify the performance of intersections, we use a combination of two performance measures from the performance measurement literature, namely split failures and ratio of arrivals on green to arrivals on red (AoR/AoG).

Split failures can be of two types: max-outs and force-offs. When a signal reaches the maximum allocated green time due to high demand, a max-out is said to have occurred, whereas, force-offs occur when an intersection reaches the maximum allocated green time without being able to fulfill the demand. High split failures are indicators of high demand on a particular phase (movement direction) of the intersection.

Arrivals on green and arrivals on red are the number of vehicles arriving at an intersection during a green phase vs. a red phase. Higher AoG are a positive indicator for signal coordination.

To conclude, intersections with high arrivals on green and low split failures are, in general, well timed and utilized. Whereas, a large number of split failures and a low AoR to AoG ratio generally indicates congestion.

We now explore these measures in more detail. A split failure is, almost always, an indicator of high demand. It can also indicate a situation where the demand is near or over the capacity of the phase (assuming no major timing problems). Most of the time, sets of intersections on the same corridor are coordinated to maximize the flow of traffic on the major street. In such a situation, the phases along the primary direction (generally, major street) will always use the maximum possible green time and hence max-out (report split failures). To ensure the usefulness of the data gathered on coordinated corridors, we take into account the detector-on events right before and after a split failure is reported. This is similar to computing the Red Occupancy Ratio (ROR) and the Green Occupancy ratio (GOR), metrics used both in the literature and in the field. Henceforth, the term split failures refers only to these demand-based split-failures.

The first step is to aggregate the high resolution (10 Hz) data into minute-by-minute buckets. For split failures reported on a phase, we record a 1 if that phase fails during the minute under consideration. More than one split failure in a minute is also recorded as a 1. If there are no split failures reported for the phase, we record a

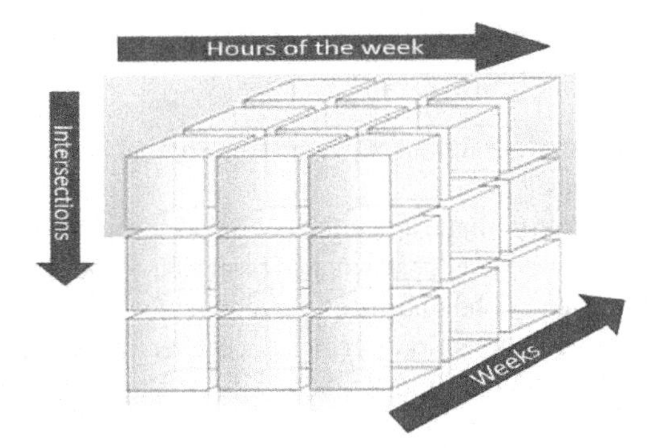

Figure 8.1: The raw data can be decomposed over three distinct dimensions i.e., intersections, hours of the week, and weeks. This can enable the discovery of both spatial (corridors of intersections that perform similarly) and temporal patterns (days of week or weeks with similar performance).

value of 0. We ignore the split failures reported in some coordinated corridors when there is no demand (ROR/GOR). Hence, we compute the duration of split failures in minutes and store that information as a time series vector. Similarly, we compute the arrivals on red, arrivals on green and the ratio of the two, every minute.

The output of the first step is two vectors with 24×60 (1,440) entries representing the behavior of a particular phase over the entire day. The next step in our methodology is to take these 1,440-digit feature vectors and aggregate into one-hour bins (we experimented with 30 min, 60 min, and 120 min bins). The dimensionality of the new vectors is 24, i.e., each vector has 24 entries representing one intersection for each the hour of the day. We now create 24×7 or 168-dimensional vectors for each intersection, representing the intersection for the whole week.

After the aggregation, we combine the vectors representing the various phases of an intersection. In our analysis, we have considered only the primary directions (phases 2 and 6) while creating these vectors. The end result is a (168-dimension) vector representing the performance of an intersection for the whole week. These 168-dimension vectors with (raw) performance measures are called fav1 (SF) and fav2 (AOR/AOG) in the rest of the document. Figure 8.1 represents a three dimensional way of organizing many such vectors. This is the vector space subsequently studied from compression and clustering perspectives. This step concludes the first stage of processing the controller log data. Thus far, we have summarized the split failures and laid the foundation for further processing.

8.1.2 COMPRESSION AND DIMENSIONALITY REDUCTION

Following the steps described in the previous section, the high resolution controller data are now abstracted into two measures that are indicative of the performance or behavior of the intersection. These are (i) split failures and (ii) arrivals on red by

green. The goal of data representation and dimensionality reduction is to enable the following:

- Automatically discovery intersections that are spatially co-located and report similar performance.
- Find daily patterns in demand and intersection performance and answer questions like whether weekday performance is similar to weekend performance or whether they need to be optimized for separately.
- Detect changes in intersection performance over many weeks and hence provide suggestions on the frequency signal timing plan updates.

The HOSVD [131] and SVD are standard approaches to linear dimensionality reduction (and they aid in denoising as well). Utilizing these tools, we find a suitable lower dimensional vector space representation of these intersection performance (fav) vectors. The aim is to find a lower dimensional representation of the data that reliably captures the performance of the intersection that is represented by these measures.

The original data is arranged in a 3D cube as shown in Figure 8.1. We use this structure for the HOSVD decomposition which employs the SVD using standard flattening of dimensions. The three-dimensional HOSVD is described in Algorithm 1. It repeatedly uses the SVD on the three unfolded, or flattened, matrices extracted from the 3D tensor.

Algorithm 1 Algorithm for Higher-Order SVD or HOSVD decomposition.

function HOSVD(A_d)
Require: A_d- High dimensional tensor.
  for $k = 1, 2, \dots d$, **do**
    Construct the standard factor-k flattening A_k.
    Compute the (compact) singular value decomposition $A_k = U_k \, \Sigma_k \, V_k^T$
    Store the left singular vectors U_k
    Compute the core tensor S. Here, $S = (U_1^H, (U_2^H, \dots, (U_d^H).A$
  end for
end function

The HOSVD uses the SVD on multiple unfoldings to obtain the core sparse tensor. We determine the best unfolding (the one that uses the least number of components) to reduce the dimensionality of the raw data while keeping the reconstruction error to a minimum. Figure 8.2 is a plot of raw data organized according to the most efficient (best) unfolding. In contrast with Figure 8.1, each row of this matrix (Figure 8.2) represents a fav vector of size 168. Many such vectors for 300 intersections over a six-month period are put together along the y-axis. Figure 8.3 is a reconstruction of the data presented in Figure 8.2 using only the top 10 components or basis vectors. In other words, we have a 10-dimensional representation of the original data.

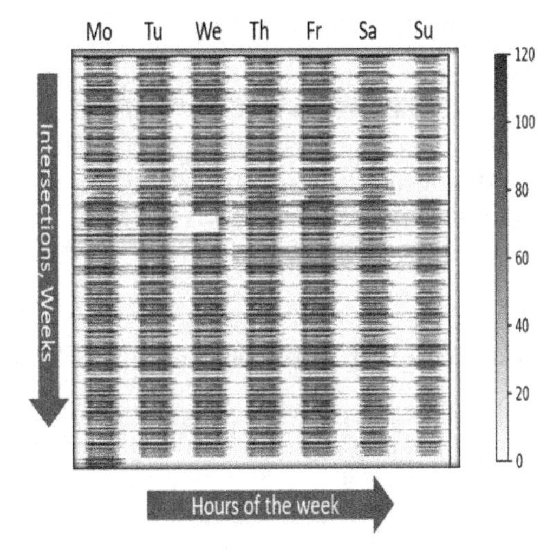

Figure 8.2: Raw representation of split failures over a six month period. Each row in this table represents a week of data for a single intersection (fav vector). This data was aggregated using the technique highlighted in Section 8.1.

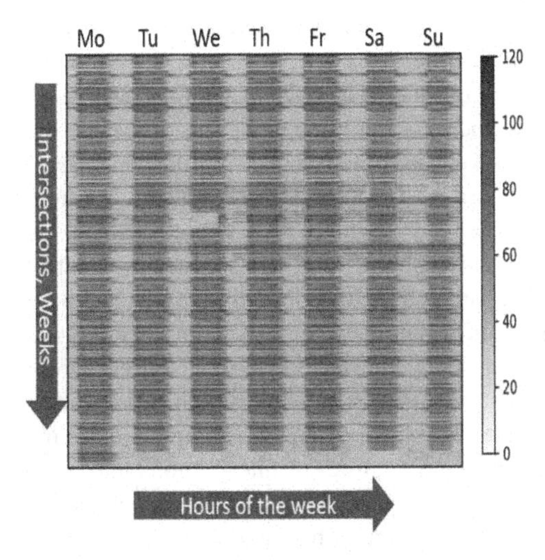

Figure 8.3: Reconstructed matrix of split failures using top 10 singular values and the selected unfolding. This can be compared with the original matrix in Figure 8.2. The original 168-dimensional data is now reduced to a 10-dimensional space.

8.2 NONLINEAR DIMENSIONALITY REDUCTION AND CLUSTERING

Linear dimensionality reduction and HOSVD (or flattened SVD) decompositions are aimed at data denoising and compression. In order to find similar patterns of traffic across space, time (hours of a day and days of the week) and intersections, we resort to nonlinear dimensionality reduction and clustering. The basic goal is the discovery of regularities across these dimensions. Because the road network geometry is known to us, we can also use this domain knowledge to find regularities in signal performance along corridors of interest.

8.2.1 DIMENSIONALITY REDUCTION AND CLUSTERING APPROACHES

Spectral clustering. Nonlinear dimensionality reduction (NLD) [7] and spectral clustering are closely related if Laplacian eigenmaps are used for NLD. A weighted graph is first constructed from the data cube with the binary relations in the graph set based on distances between the vectors. Since we already have a linear vector space embedding of the raw data (from the previous section), this metric is used as the distance measure. Nonlinear dimensionality reduction proceeds by computing the generalized eigenvectors corresponding to a suitable (usually the lowest) set of eigenvalues of the graph Laplacian. The rows of the eigenvector matrix form the nonlinear embedding. Given this embedding structure, we can then perform standard clustering in this space (K-means or equivalent) to produce the clusters of interest. If necessary, the tight relationship between spectral clustering and graph partitioning can be exploited to obtain a different set of clusters, but we have found that in practice standard K-means clustering in the embedded space works well.

Model selection, number of clusters. Using the Elbow method, we plot the change in the sum of squares distance between all the data points and their assigned cluster centers as we increase the number of clusters. This reduction in SSE (Sum of Squares Error) is shown in Figure 8.5. From the plot, we can conclude that the number of clusters K must be at least 5. However, this method does not give us an upper bound on K. After $K = 5$, the reduction in SSE is mostly linear. After trying various values above 5, we picked $K = 10$ as this allows us to order the clusters by the performance observed at the cluster centroid. This can be seen in Figure 8.4.

Spatial information. Sometimes the intersections belonging to a cluster are spread over geographic regions many miles apart. While these intersections may be performing similarly, there is no real value in having such distant intersections in the same cluster if we want to modify signal plans, for example. So, we may do a second round of processing where we split a cluster of intersections into multiple disjoint clusters based on a geographical indicator like primary road names, distance, or the hop distance between the intersections. This is done in Section 8.3 for change detection only.

8.2.2 ORDERING CLUSTERS BY PERFORMANCE

In the previous section, every intersection for each week is assigned to one of 10 clusters (Figure 8.4) based on the observed split failures. Given these clusters, we

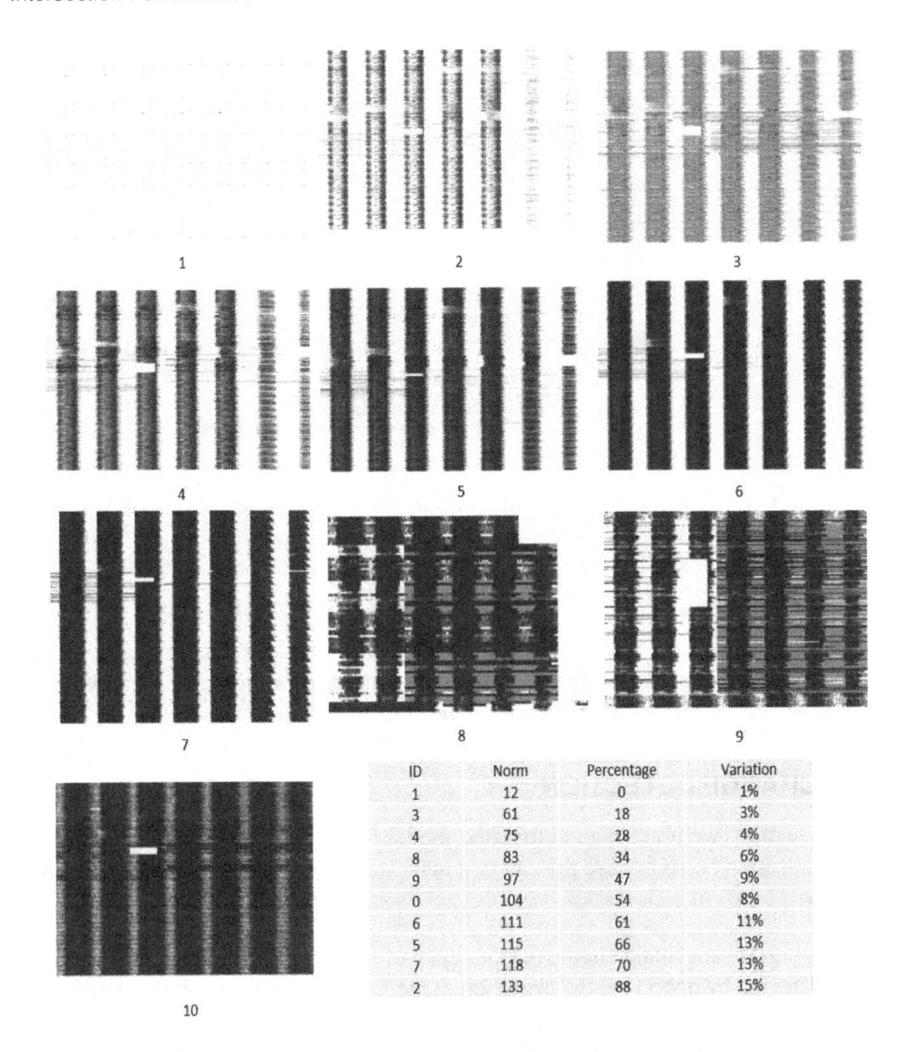

ID	Norm	Percentage	Variation
1	12	0	1%
3	61	18	3%
4	75	28	4%
8	83	34	6%
9	97	47	9%
0	104	54	8%
6	111	61	11%
5	115	66	13%
7	118	70	13%
2	133	88	15%

Figure 8.4: The 10 clusters corresponding to distinct demand patterns discovered in the data. We order these clusters from low demand to very high demand using the norm values of the cluster centers presented in the corresponding table. This is derived using the techniques presented in Section 8.2.2 which give us a partial ordering of the clusters.

now compute the 1-norm of the fav1 vectors for each cluster centroid. In this special case, because of the clear differences in performance of intersections, we get a linear ordering of the clusters in terms of intersection performance objectives. Note that in this case, the 1-norm can be interpreted as number of minutes of observed split failures at the intersection. Using this information, we can reorder the clusters in increasing order of observed split failures (or perceived performance). This can be

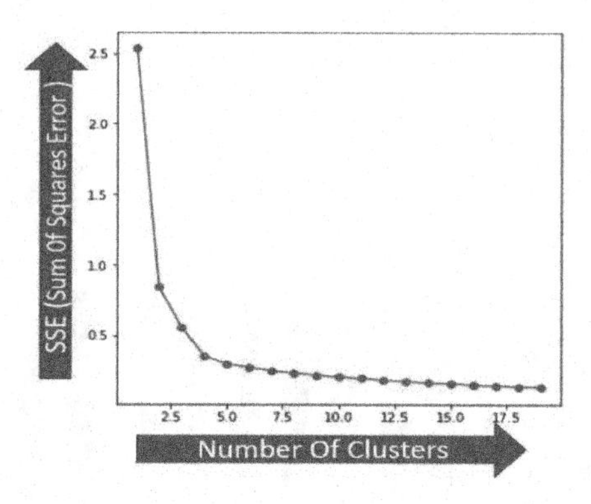

Figure 8.5: This is a plot of the reduction in the sum of squares distance between all the data points and their assigned cluster centers as we increase the number of clusters K. From this plot, it can be concluded that K must be chosen to be at-least 5.

clearly seen in Figure 8.4. The figure also includes a table of the computed norms and the linear ordering. We also compute the 1-norm of all the observations in the cluster to get the percentage variation in the cluster.

8.3 CHANGE DETECTION

In this section, we present an automated model for detecting significant changes in the intersection behavior. We present two main ways to accomplish change detection :

- A significant change in the performance of the intersection over time can be detected by observing the evolution of the intersection's cluster membership over time.
- A change in the lower dimensional projection of the data representing the intersection performance over weeks can be used as a change detection measure.

In practice, the two methods are combined into one overarching method. In our case, we find it convenient to examine the changes in the representation of each cluster. It turns out this is most convenient to perform in the SVD basis of the intersections. We briefly describe the approach. The SVD bases of all intersections are first assembled. Subsequently, we compute the distances between the basis vectors. If these distances increase out of proportion to the "normal" distances, that could herald a change. In addition, we can also examine the changes in the cluster membership of each intersection. If this suddenly changes, that could be significant. The change in cluster membership can also be exploited in terms of intersection performance: whether it improves or worsens can be recorded. Figure 8.6 depicts these changes in

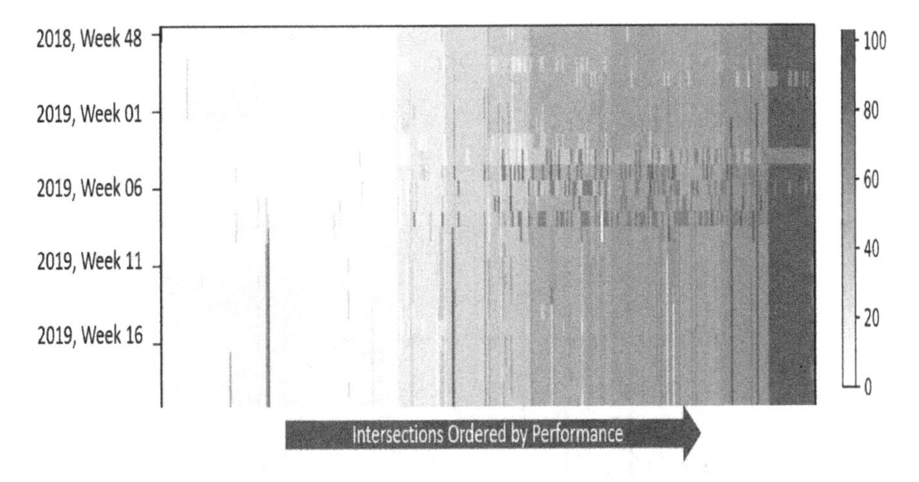

Figure 8.6: A performance based representation of the overall dataset. Each pixel in this plot represents an intersection for an entire week. The color, visible in our ebook version, is based on the cluster center to which the observation (intersection, week) belongs. It represents the percentage of split failures observed at the cluster center. The rows represent a period of 25 weeks (~6 months).

performance. In the figure, each pixel represents an intersection for an entire week. This is in contrast with the previous plots. The color used is based on the cluster center for the observation and represents the percentage of split failures observed at the cluster center. The rows represent a period of 25 weeks (~6 months). Similarly, in Figure 8.7, the rows correspond to weeks, and the columns correspond to intersections. Each pixel still corresponds to aggregate performance of the corresponding cluster (in terms of percentage of split failures) for one week. Further, the columns are sorted so that all intersections within a corridor are adjacent to each other, and the corridors are sorted based on their congestion behavior in increasing order. The best performing corridors are to the left. We also show temporary temporal changes at a single corridor and at multiple corridors.

Local change detection and retiming recommendations. In Figure 8.8, (a) split failures and (b) the arrivals on red/green ratio (AoR/AoG) are plotted for a single corridor. Again, the rows correspond to weeks, and the columns correspond to intersections. For split failures, each pixel corresponds to aggregate performance of the corresponding cluster (in terms of percentage of split failures). And the AoR/AoG plot represents the aggregate ratio observed at the intersection for that week. These two charts in combination allow us to highlight parts of corridors that are performing well and the ones that are in need of immediate attention.

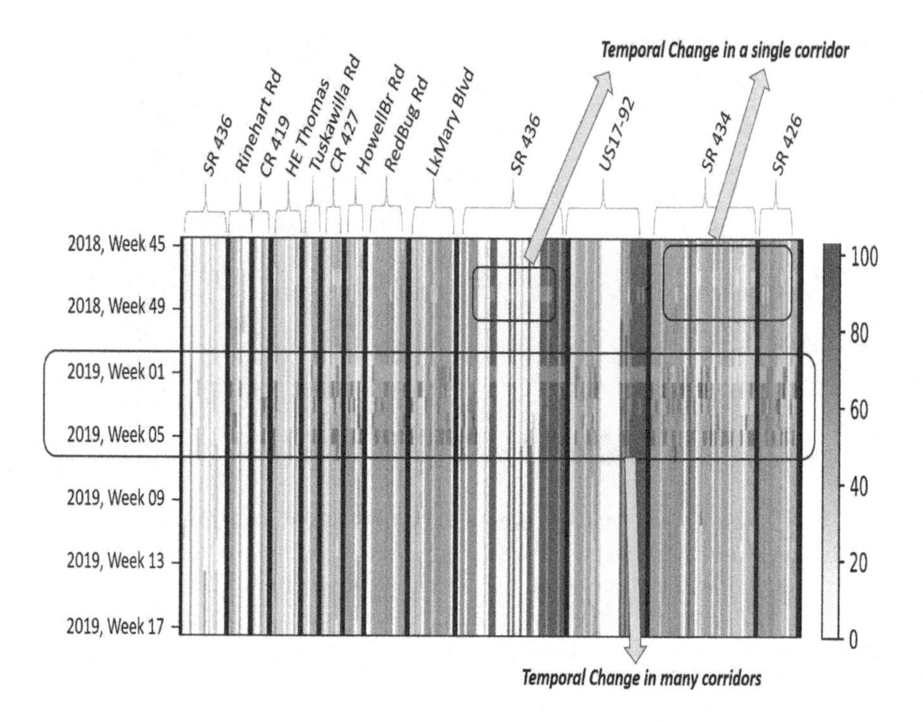

Figure 8.7: A spatiotemporal representation of the overall dataset. The rows correspond to weeks and the columns correspond to intersections. Each pixel corresponds to aggregate performance of the corresponding cluster (in terms of percentage of split failures) for one week. Further, the columns are sorted so that all intersections within a corridor are adjacent to each other, and the corridors are sorted based on their congestion behavior in increasing order. The best performing corridors are to the left. We also show temporary temporal changes at a single corridor and in multiple corridors.

8.4 CONCLUSION

In this chapter, we have presented a data-driven approach to process high resolution intersection controller logs (with the eventual goal of building a decision support system). We demonstrated this approach using split failures as the primary MoE to categorize and classify intersections according to their performance, automatically detect changes in performance, and highlight intersections and regions that may need immediate attention. This approach can be used to quickly identify the most problematic intersection in the region.

To accomplish this, we used a pipeline of denoising, dimensionality reduction, and spectral clustering techniques on the aforementioned split failure and AoR/AoG features to group together signals exhibiting similar behavior. Next, we used the homogeneity of the obtained clusters to re-order these clusters in terms of performance, and

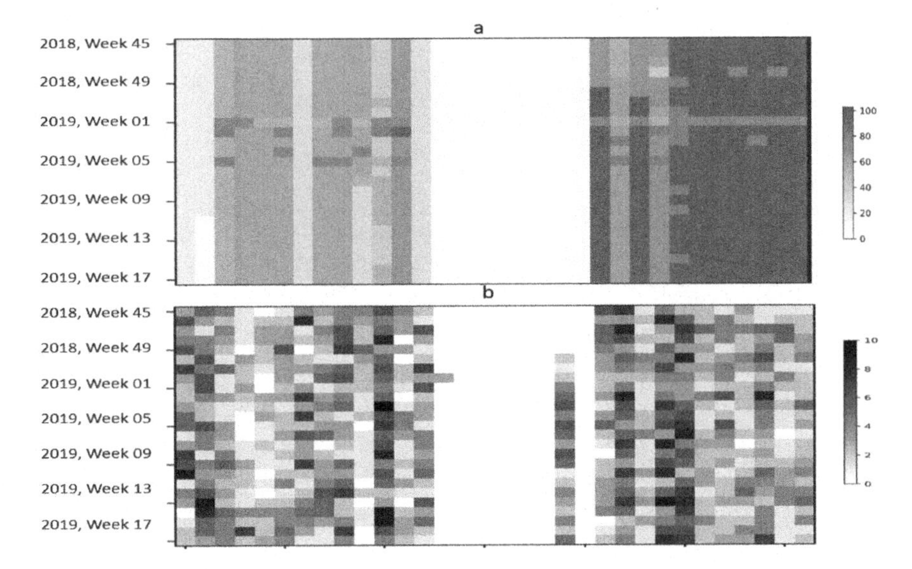

Figure 8.8: We look at a single road or corridor from the previous figure, in this case, US17-92. We can see the percentage of split failures (of the cluster center) on the top in (a) and the raw value of the AoR/AoG ratio in the bottom (b).

hence, automatically highlight the change or evolution of intersection performance with time. This also enables the separation of changes that result in an improvement of the observed performance from those that result in poorer intersection performance.

These results can be used to proactively identify problematic intersections. Once the problematic intersections are known using our (and related) techniques, the ATSPM system can be queried in order to drill down and analyze the intersections. Hence, our techniques can complement existing ATSPM systems by providing summarization and change detection information in a human-understandable manner.

9 Interruption Detection

Abstract - Detection of traffic interruptions is a critical aspect of managing traffic on urban road networks. This chapter outlines a semi-supervised strategy to automatically detect traffic interruptions occurring on arteries using high resolution data from widely deployed inductive loop detectors. The techniques highlighted in this paper are tested on data collected from detectors installed on more than 300 signalized intersections over a 6 month period, hence can detect interruptions with high precision and recall.

Managing traffic incidents is one of the crucial activities for any traffic management center. These incidents are non-recurrent, may arise due to different causes like traffic accidents, vehicle breakdowns, debris etc. and cause congestion.

It is worth noting that not all accidents (e.g. a fender bender) result in interruptions. From a traffic management perspective, it is more important to detection significant interruptions rather than accidents. Further, this should be done in real-time so that proactive actions can be used for mitigation. We define an interruption to be time periods where the amount of traffic is significantly lower than normal traffic and for a significant period of time. The focus of this study is in detecting these events of interest from inductive loop detectors installed at signalized intersections.

There has been significant related work in incident detection using social media data (e.g. Waze) and probe-based systems (e.g. GPS or Bluetooth tracking data). TMCs have had limited operational use of automatic incident detection techniques because of high rates of false alarms and low detection rates [148]. In fact, many automatic incident detection algorithms perform poorly in the real-world, compared to simulated traffic environments [109] (and please see Section 9.1 for more details). Also previous work using loop detector data is generally limited to simulation or small datasets.

In this chapter, we describe how to use large scale loop detector data for detecting traffic interruptions. With the advent of new systems, loop detector data is available at high frequency (10 Hz) and with low latency. Hence, the utilization of this data for determining traffic interruptions can have wide applicability and can be used in conjunction with other systems based on human reporting or with probe-based systems.

Labeling data is a major challenge for this application since ground truth is generally not available. We define traffic interruption as a significant, contextual and non recurring change observed in a combination of the following parameters: the amount of deviation of traffic volumes from predicted volumes and the duration for which the actual traffic volume deviated from the predicted volume. This definition is then used for automatically labeling events of interest (EOI) from historical data.

Our algorithms use time-series data for detecting if an event of interest has occurred using traffic information from recent and historical data (from similar time periods on previous days or weeks) to predict if an event of interest—defined as a long traffic interruption—has occurred. Whether or not an EOI has occurred depends on a key parameter—the duration of time after reduction in traffic at a single detector. This has an impact on the overall accuracy (in terms of false positives and false negatives). In particular, we find that waiting 60 to 90 seconds after a significant reduction in traffic is reasonable to determine EOIs with high accuracy and low latency.

We use fine grain (10Hz) data for 300 intersections over 6 months to demonstrate the usefulness of our method. Experimental results show that our approach has a high degree of accuracy.

The rest of the chapter is outlined as follows. The first step in this process is to analyze and pre-process raw data gathered from fixed point sensors or detectors. This is described in Section 9.2. Once the data has been processed, the next step in the pipeline is quantifying what an interruption is and also labelling said traffic interruptions. This is described in Section 9.3. The third step is to develop algorithms for predicting the labeled interruptions. This is described in Section 9.4. Conclusions are provided in Section 9.5

9.1 RELATED WORK

The existing literature pertaining to incident detection can be broadly classified as follows:

1. Traditional systems which rely on inductive loops and video cameras for vehicle detection.
2. Probe-based systems (GPS data from fleets of vehicles like NavTech or HERE data).
3. Human reporting systems like calls to traffic management centers or the use of social media platforms (like twitter).

There is also some work on using a combination of multiple datasets.

Most of the existing research in automatic incident detection is focused on freeways or uses simulated data. The basic idea behind these approaches is that if an incident occurs, there would be a significant decrease in the occupancy at the downstream detectors and increase in occupancy at upstream detectors [2, 18, 21, 53, 60, 63, 77, 84, 128, 129]. Incidents are detected by comparing current occupancy or speed value with the derived thresholds. These methods are not directly applicable to target application as urban road networks have a high density of signalized intersections an they behave differently from freeways due to influence of traffic signals, pedestrian crossings etc.

There is an extensive body of research [4, 100, 108, 120, 158] on incident detection using probe based systems. The advantage of probe data over fixed detector data is that probe data cover longer sections of the road which can also be used to detect secondary incidents [108, 158]. These approaches depend on the penetration rate of the probe car and confidence level of the data. Additionally, many of these datasets are useful for computing travel time and not necessary for traffic interruption. The former can increase/decrease based on the level of congestion and/or signal timing. Additionally, this data is generally sold by companies and can be expensive.

Algorithms based on human reporting systems make use of sources like Twitter, phone calls, Waze etc. In [49], the authors presented methods to mine tweet texts and extract information related to incidents.

Other related research on incident detection on arterials [2, 18, 90] relies on simulated data (and accidents) or assumes the availability of ground truth (either using simulations or labeling). Due to this, many automatic incident detection algorithms perform poorly in real world scenarios when compared to simulated environment [109]. Moreover, developing an incident data-set with start and end times can be tedious and requires manual investigation by TMC personnel.

Taking into account the issues highlighted above, this chapter focuses on detecting traffic interruptions based on real, fixed point sensor data (detector data) collected from signalized intersections and detectors on urban road networks. In the next section, we focus on the data processing needed for near real-time incident detection. Due to the real world focus, we believe that the results presented in this paper can be translated into practice.

9.2 DATA PREPROCESSING

The high resolution data was processed based on cycle length (time between start of successive start of green phases). For 8 am to 8 pm, the intervals correspond to a cycle or a fixed size interval. For 8 pm to 8 am, the interval size is fixed to the median of cycle lengths from 8 am to 8 pm (because cycle lengths during the overnight period are significantly large and erratic). A cycle-based approach is expected to be more predictive as it maintains the consistency of phases over time. Table 9.1 shows a sample of a processed time series data with the following attributes: Timestamp (timing pattern start time), Count (number of arrivals in this cycle), and CL (cycle length in seconds). We removed intervals of data where detectors are broken/not reporting any data for significant amount of time on some days. We removed intervals of data where the cycle length is less than a second. We use this time series data to quantify, label traffic interruptions section 9.3. Python multiprocessing packages and Pandas Python data processing library are used to process data for several intersections simultaneously in a multicore machine. Time requirement is approximately 5 min for one week of data, 300 intersections on a machine with 54 cores, and 256-GB memory. This suggests that the computational requirements are sufficient for implementation in near real-time scenarios.

9.3 LABELLING INTERRUPTIONS

To be able to detect traffic interruptions we need a robust mechanism for labelling them, the labelling mechanism should be separate from the detection algorithm. In this section we describe our methodology for labelling by quantify traffic interruptions at a detector level(single lane), approach level(multiple lanes). For this we define an interruption using two parameters as follows

Table 9.1

Processed representation of the raw data from Figure 1. Different attributes include (a) *Timestamp*: start of green time, (b) *Count*: number of arrivals in each cycle, (c) *CL*: cycle length in seconds, (d) *AR*: arrival rate, (e) *D-ID*: detector ID, (f) *S-ID*: signal, ID, and (g) *Ph*: phase.

Timestamp	Count	CL	AR	D-ID	S-ID	A-ID	Ph
2018-08-01 00:00:00	5.0	210.0	0.0238	149016	1490	8612	6
2018-08-01 00:00:00	6.0	210.0	0.0285	149006	1490	8611	2
2018-08-01 00:00:00	4.0	210.0	0.0190	149007	1490	8611	2
2018-08-01 00:00:00	6.0	210.0	0.0285	149008	1490	8611	2
2018-08-01 00:00:00	7.0	210.0	0.0333	149018	1490	8612	6
2018-08-01 00:00:00	8.0	210.0	0.0380	149017	1490	8612	6
2018-08-01 00:00:00	6.0	210.0	0.0285	149009	1490	8611	2
2018-08-01 00:03:30	9.0	210.0	0.0428	149018	1490	8612	6
2018-08-01 00:03:30	9.0	210.0	0.0428	149006	1490	8611	2
2018-08-01 00:03:30	2.0	210.0	0.0095	149016	1490	8612	6
2018-08-01 00:03:30	5.0	210.0	0.0238	149017	1490	8612	6

- The amount of deviation in terms of percentage reduction in arrival volumes from the baseline arrival volumes.
- The duration(in seconds) for which the actual arrival volumes is less than the baseline volumes.

For our baseline prediction we look at arrival volumes (in recent cycles) and historical arrival volumes from similar time periods(same time of the day and day of the week). We found that a simple baseline predictor works well in practice and that our approach is not terribly sensitive to the choice of method. Arrival volumes in any

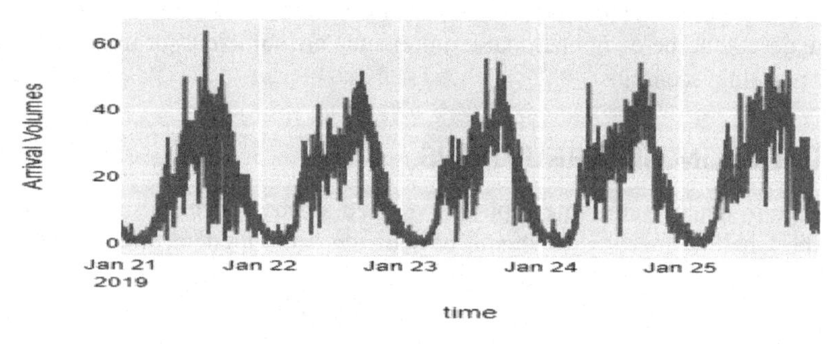

Figure 9.1: Arrival volumes visualized as time series for a detector for six days. It is worth noting the cyclic nature of data.

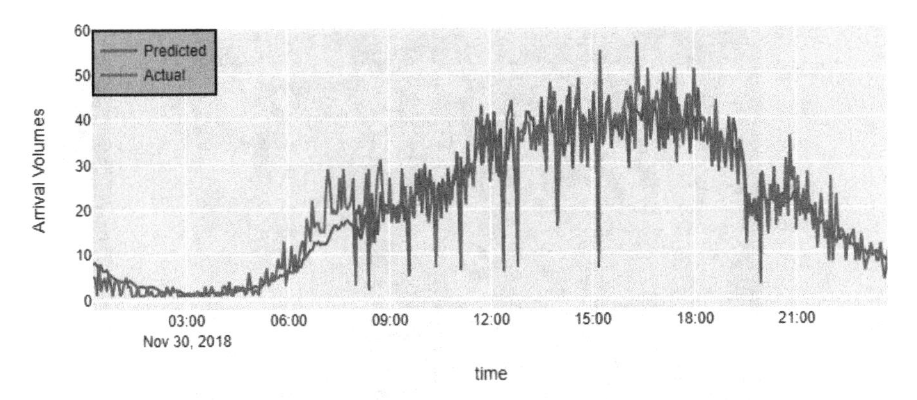

Figure 9.2: Actual and predicted volumes vs time for a period of 24 hours, and a single detector showing predicted volumes are largely consistent with actual traffic patterns.

cycle is dependent on length of the cycle, so we regress on arrival rates instead. Let V_i and T_i correspond to the number of arrivals and the duration of cycle i respectively. Then the arrival rate, X_i, is defined as $X_i = \frac{V_i}{T_i}$. Our model for f is

$$X_t = f(X_{t-1}, X_{t-2}, \ldots, X_{t-k}, Y_{t-k}, \ldots, Y_{t+k}),$$

where X_i and Y_i corresponds to arrival rates from the current day and historical data respectively. We use this function to compute arrival volumes at any given cycle using historical data, label interruptions in terms of deviation from this baseline volumes.

Figure 9.2 shows actual vs predicted arrival volumes at a single detector. This plot suggests that our baseline prediction mechanism is good enough as predicted volumes are highly consistent with actual traffic patterns. *.An Event of Interest $EOI(Y,T)$ will refer to and with Y% reduction for T seconds.* The 2D histogram, Figure 9.3, shows bi-variate distribution of EOI based on percentage volume reduction(Y), duration(T). We are interested only in interruptions that are long enough, volume reduced by significant amount to require addressing and also relatively infrequent (discarding normal behaviour).

Each detector corresponds to a single lane. This can be extended to approaches with multiple lanes as follows. $EOIA(Y,T)$ - Y corresponds to the fact that after an interruption as defined earlier on any of the lanes, the minimum reduction of the traffic for all the lanes and T is the minimum duration of the interruption along any of the lanes. Thus, if one lane is not interrupted, the corresponding value of Y and T may be zero.

In the next section we describe our methodology for detecting interruptions at a single detector level as well approach level.

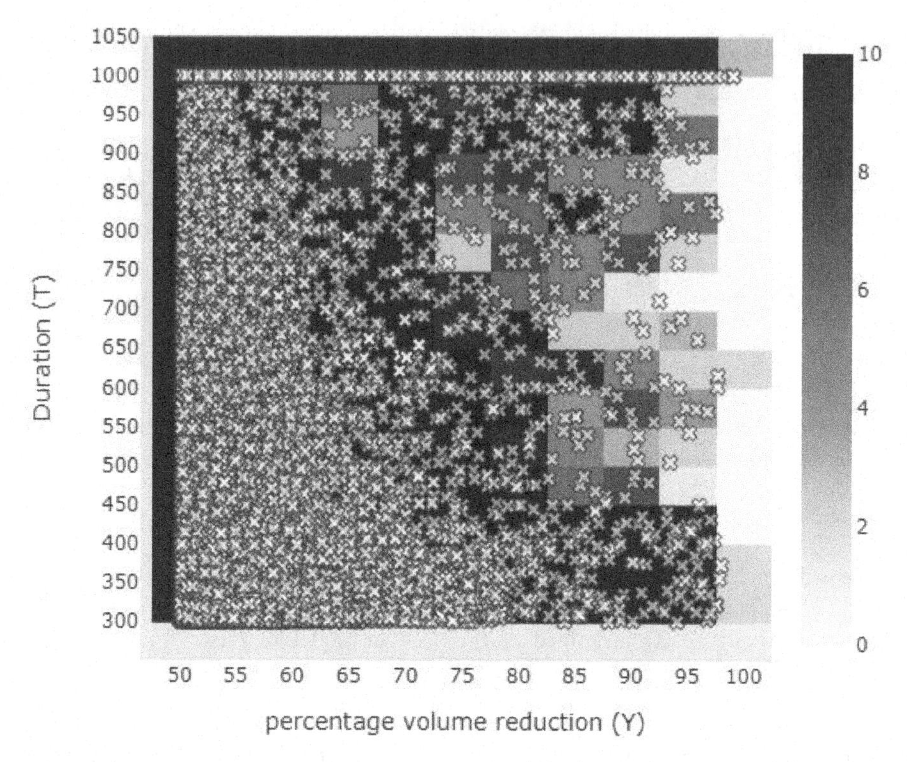

Figure 9.3: Two dimensional histogram showing bivariate distribution of interruptions on a single detector based on reduction in volume percentage and duration in seconds

9.4 DETECTING INTERRUPTIONS

We design a machine learning for detecting Events of Interest (EOI) assuming that at any given point we have historical data available for a similar time period and recent data for that detector or approach. The first step of this process is to limit cycles if the current volume is smaller than the predicted volume. If this condition is met, we say that a trigger has occurred and we compare the current cumulative volumes with historically relevant cumulative volumes. we construct the following cumulative curves:

- The red curve, starting from the beginning of the previous cycle as shown in Figure 9.5.
- The curves in green are cumulative arrival curves for the same time interval as in curve 1 from cycles based on historical data from the same time of the day and day of the week.

Current cumulative curve concatenated with historical cumulative curves is treated as input to the classifier. The output for a binary classifier that we learn is whether the current cumulative distribution belongs to EOI or not.

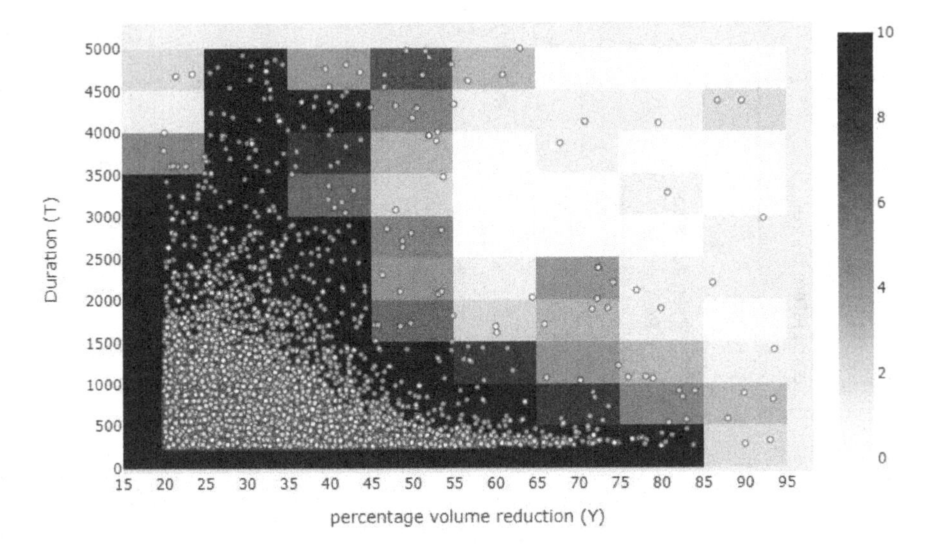

Figure 9.4: Two dimensional histogram showing bivariate distribution of interruptions on two lane approaches based on reduction in volume percentage and duration in seconds

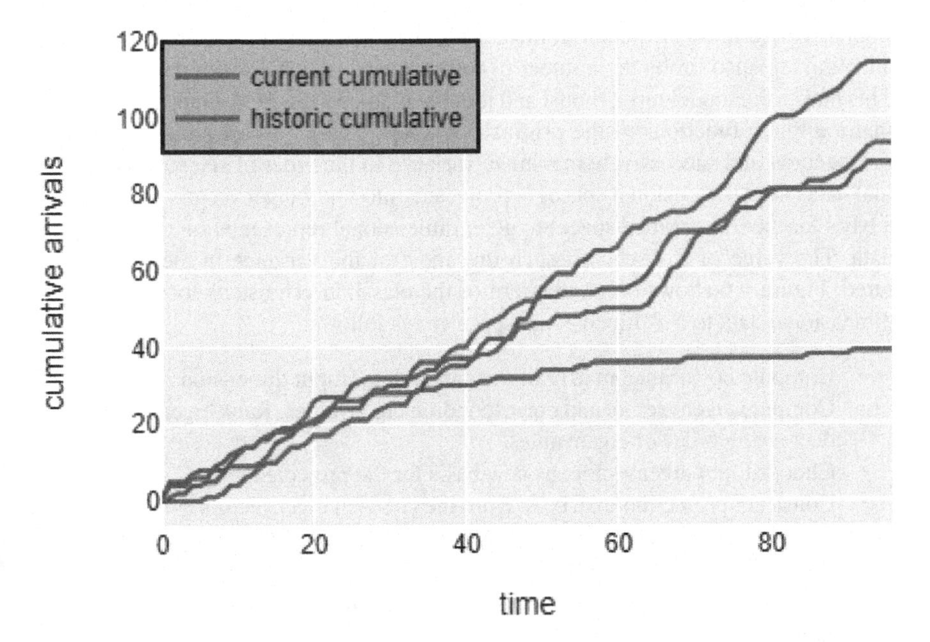

Figure 9.5: Plot showing current and historical cumulative curves. In this case, it shows that there is potential event of interest. The more the lag in decision making the higher the probability of making a correct decision.

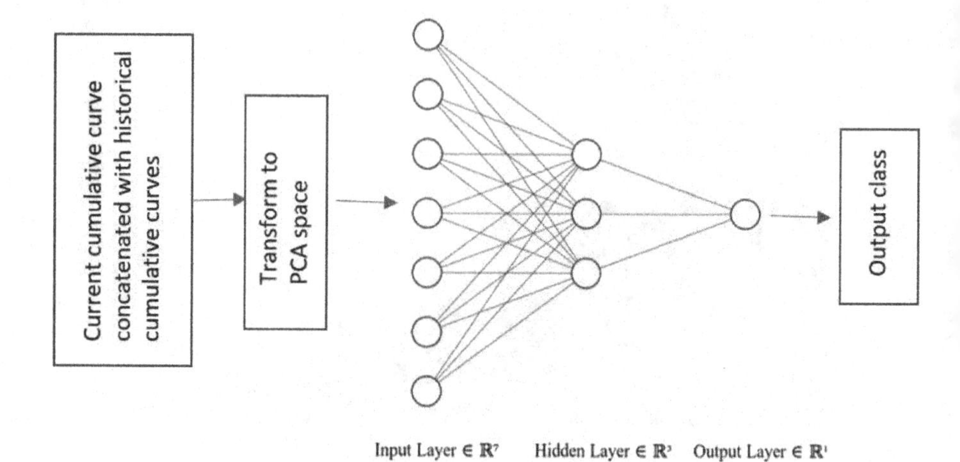

Figure 9.6: Block diagram showing architecture of a three layer classifier. The input is first transformed into a lower dimensional space using PCA.

We first transform the input time series to a low dimensional subspace, using (PCA), where most of the variability of the data is captured. This improves generalization and also limits the number of weights used in our discrimination network. This makes the model more robust and less likely to over-fit. PCA finds new variables that are linear functions of the original variables. These new variables are linearly independent and successively maximize variance in the order of selection. The original data (in D dimensional space) is projected onto top d eigenvectors (which act as a basis for the transformed space) to get d dimensional representation of the original data. The value of d is chosen such that most of the variance in the data is captured. Figure 9.6 shows block diagram of the classifier. The steps for projecting D dimensional data to a d dimensional space are as follows:

- Compute covariance matrix of the data points (Input dimension D).
- Compute eigenvectors and corresponding eigenvalues. Rank eigenvectors in decreasing order of eigenvalues.
- Choose top n eigenvectors as new basis for the projected space.
- Construct projection matrix W from the choosen eigenvectors
- Use projection matrix to transform original data in D dimensional space to new d dimensional subspace.
  Transform the original D dimensional data points to this new d dimensional space.

Figure 9.7 shows plot of cumulative explained variance vs number of components when PCA is applied to the input time series. This plot suggests that taking top 20 components captures 99% of the variance of the data.

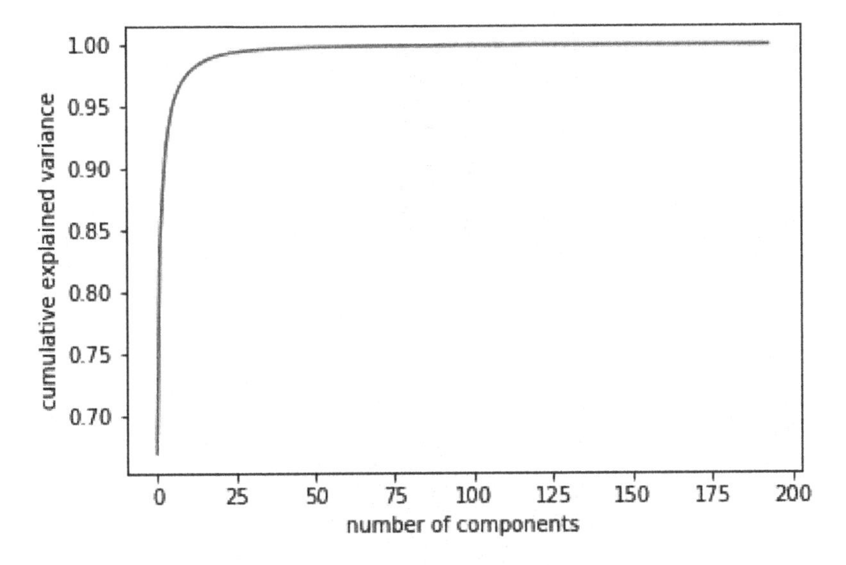

Figure 9.7: Plot of cumulative explained variance vs. no. of components. This plot suggests that taking the top 20 components captures 99% of the variance of the data.

Table 9.2 provides the precision recall for a variety of EOIAs based on three parameters - percentage reduction in volume, duration and wait time. EOI(Y> 60, T> 400) is the set of EOIs with at least 60% reduction for at least 400 seconds. These results show that our neural network approach is highly accurate. For EOI(Y>70,T>500)

Table 9.2

Precision and Recall for different EOI

Classification task	EOI	Wait Time	Precision	Recall
Approach - Two lane	EOIA (Y>60,T>500)	90 sec	80	79
Approach - Two lane	EOIA (Y>50,T>400)	90 sec	87	82
Approach - Two lane	EOIA (Y>40,T>400)	90 sec	85	83
Approach - Two lane	EOIA (Y>30,T>400)	90 sec	83	82
Approach - Two lane	EOIA (Y>30,T>300)	90 sec	66	66
Approach - Three lane	EOIA (Y>50,T>400)	90 sec	96	98
Approach - Three lane	EOIA (Y>40,T>400)	90 sec	70	91
Approach - Three lane	EOIA (Y>30,T>400)	90 sec	70	84
Detector level	EOI (Y>70,T>500)	90 sec	98	97
Detector level	EOI (Y>70,T>500)	60 sec	87	83

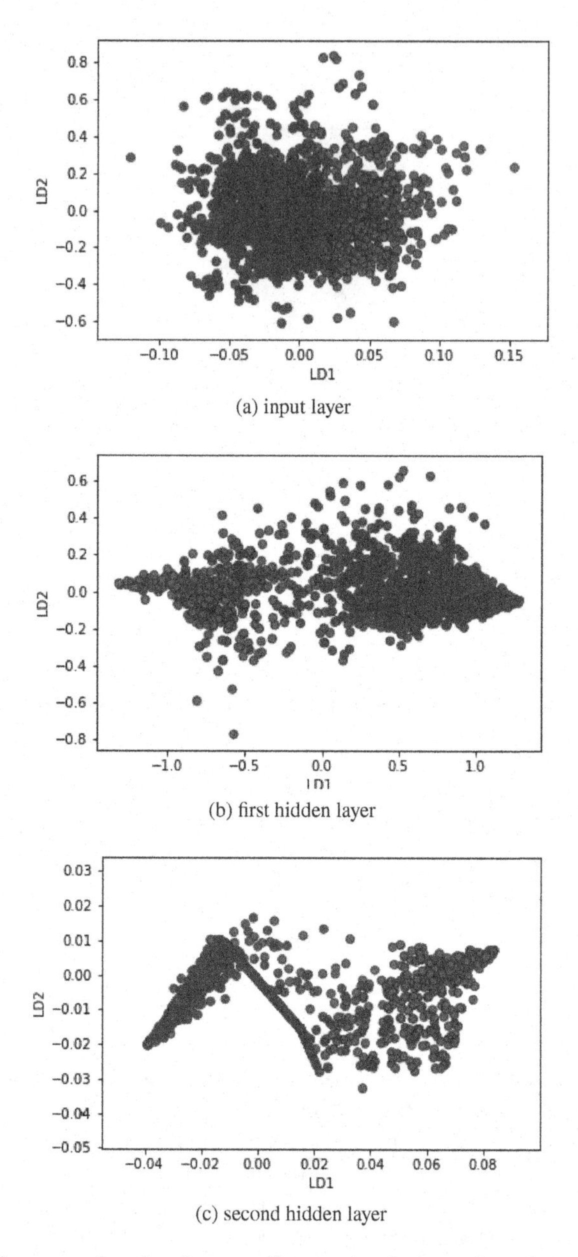

(a) input layer

(b) first hidden layer

(c) second hidden layer

Figure 9.8: Diagrams showing intermediate activations projected on to two dimensional fisher space at different layers. The positive and negative cases are displayed in two distinct colors in our ebook version. These diagrams show that the two classes seem more separable at the higher levels

with a 90 second lag after trigger to make a decision, we have an accuracy of over 98% in correctly classifying EOIs for a single detector. For a three lane approach our accuracy is over 87%.

The precision, in general, increases as the lag increases. For example for 60 seconds to 90 seconds precision increases from 87% to 98% for EOI (Y>70, T>500) at a single detector level. This allows a traffic engineer to perform the appropriate trade-off

It is also worthwhile to understand the underlying behavior of the neural network on how it transforms the input time series, in making the final decision. We show the intermediate activations of the layers of the neural network by projecting these activations onto the linear fisher discriminant space. We present the maximal discriminatory features in two dimensional space (Figure 9.8). We can see that the neural network applies non linear transformations in such a way that the two classes become more separable after each layer. The output sigmoid layer of the neural net acts as linear classifier upon the activations of last hidden layer. This demonstrates the benefit of using multiple layers and non-linear activation functions.

9.5 CONCLUSIONS

In this chapter, we presented a machine learning approach for detecting traffic interruptions in the absence of labeled data. We present a systematic approach to define events of interest at the detector and approach level. We describe a framework that uses machine learning algorithms based on historical and current data. These algorithms have high precision and recall for them to be useful in practical scenarios. Additionally, this approach provide trade-offs between lag of decision and accuracy to allow an application engineer to make appropriate choices.

10 Estimating Turning Movement Counts

Abstract - Predicting intersection turning movements is an important task for urban traffic analysis, planning, and signal control. However, traffic flow dynamics in the vicinity of urban arterial intersections is a complex and nonlinear phenomenon influenced by factors such as signal timing plan, road geometry, driver behaviors, queuing, etc. In this chapter, we describe deep neural networks that are capable of directly learning an abstract representation of intersection traffic dynamics using detector actuation waveforms and signal state information. This chapter describes in detail, neural network models for predict turning movement counts with greater accuracy when higher resolution data are provided.

Turn movement counts (TMCs) are used for a wide variety of applications related to intersection analyses, intersection design, and transport planning. Applications such as signal timing optimization or generating diversion routes need granular data on TMCs.

The traditional method of using a human to obtain this information manually is very time consuming and monotonous. Although approach volumes can be easily derived from the loop detectors, they only give information on how many vehicles are entering the intersection and not the TMCs. The focus of this work in on accurately and efficiently deriving TMCs using loop detector data by utilizing machine learning techniques.

So far, intersection systems have only given data at coarse levels of granularity (for example, traffic movement counts by hour) limiting their use. The recent advent of automated traffic signal performance measures (ATSPM) systems provides this information and signal timing information at a decisecond level. Availability of this information along with the availability of cheap GPU-based computing and deep neural network algorithms has opened up the possibility of modeling traffic behavior at the subcycle level and is the focus of this chapter.

In this chapter, we present neural network models for predicting turning movement counts for an intersection, given information from stop bar detectors and advance detectors (Figure 10.1) of the intersection and advance detectors of upstream intersections. In this approach we use detector waveforms instead of aggregated traffic volumes for modeling TMCs. The former leads to better accuracy for turning movement counts prediction. These neural network models perform well both under unsaturated and oversaturated conditions.

10.1 RELATED WORK

In this section, we first summarize the existing literature on turning movement counts prediction. We broadly categorize the existing literature and summarize each category.

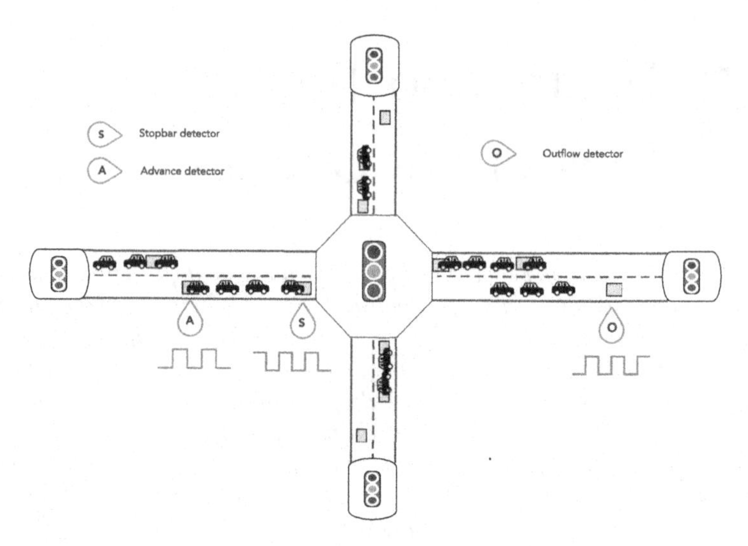

Figure 10.1: A typical four leg intersection. Arrival/departure waveforms are observed at stop bar (S) and advance (A) detectors.

10.1.1 FLOW-BASED METHODS

Some studies, [17, 106, 152, 156], have used mathematical techniques like linear programming to impute turning movement counts which balances inflow, outflow of traffic with some constraints. By modelling the flow with some constraints as linear equations, we can solve for every unknown movement if we have enough equations. To calculate westbound through volume at intersection i, the following equation can be used.

$$\begin{aligned} WBT_i^t &= -NBL_i^t - SBR_i^t + WBR_j^{t+\delta t} \\ &\quad + WBL_j^{t+\delta t} + WBT_j^{t+\delta t} - \delta_{ij}^t \end{aligned} \tag{10.1}$$

where

WBT: westbound through
NBL: northbound left
SBR: southbound right
WBR: westbound right
WBL: westbound left
t: time interval t
i,j: intersections number
Δt time taken to travel between i,j
$\delta_{ij}^{t+\Delta t}$: additional trips generated between intersections i, j during time interval Δt.

In the equation, Δt is assumed to be zero if the distance between the two intersections is small because the volume fluctuations would be significantly low. Also,

δ_{ij} can be assumed to be zero if there are no trips generated between these two intersections.

Linear programming-based approaches [106] have been used to compute turning movement counts using flows. This approach requires measurements like detector flow, weight given to each link, and constraints on flow for each link. A system of linear equations is modelled with the above constraints and information. This system of equations is solved to get turning movement counts.

Methods based on origin-destination (O-D) matrices to compute turning movement counts [152] uses already observed O-D matrices to predict future O-D matrices. They use a travel demand model, the Furness Method, and then use the equilibrium principle to assign future O-D matrices to the network.

The path flow estimator [17] computes complete link flows along with turn movement counts when some traffic counts at selected intersections are known. The proposed algorithm iterates over entropy and equilibrium terms to obtain the final solution. The first term, entropy, distributes trips to multiple paths. The second term, equilibrium, makes those trips cluster together on minimum cost paths. The algorithm has an outer loop to iteratively generate paths to the working path set as needed to replicate the observed link counts, turning movement counts, and selected prior O-D flows. The claim is that the method has advantages because of the single level convex programming formulation, when compared to other bilevel programming approaches.

10.1.2 NEURAL NETWORK BASED APPROACHES

These methods [45] use only approach volumes to predict corresponding turning movement counts. Figure 10.2 shows how a four-leg intersection is represented as a node where the number of vehicles leaving the node i is P_i: $P_i = \sum_{j=1}^{4} T_{ij}$. The turning movement counts between node i and node j are the number of trips from i to j. So, when a vehicle is approaching an intersection, they have four possible turns: left, through, right, and U-turn. For the four approaches, there are 16 possible movements. For intersection j, the number of vehicles attracted to it is A_j: $A_j = \sum_{i=1}^{4} T_{ij}$; this is the sum of all the trips with destination j. So 16 different turning movement counts should be modelled with the help of known values. Each T_{ij} is modelled as a function of P_i and A_j. [90] uses a stacked autoencoder model to predict traffic flow. An autoencoder is a neural network that tries to reproduce its input. It acts as an identity function. It tends to learn features that form a good representation of its input. Stacked autoencoder models are created by stacking autoencoders in a hierarchical way. The generic traffic flow features are modelled with a stacked autoencoder model. They also take into account spatial and temporal correlations. They use a logistic regression layer as the top layer for supervised traffic prediction.

10.1.3 OTHER APPROACHES

Approaches such as the genetic algorithm-based framework [111] are also used to obtain a dynamic relation between turning movement proportions and traffic counts at an intersection at each time step. The proposed objective function minimizes the sum

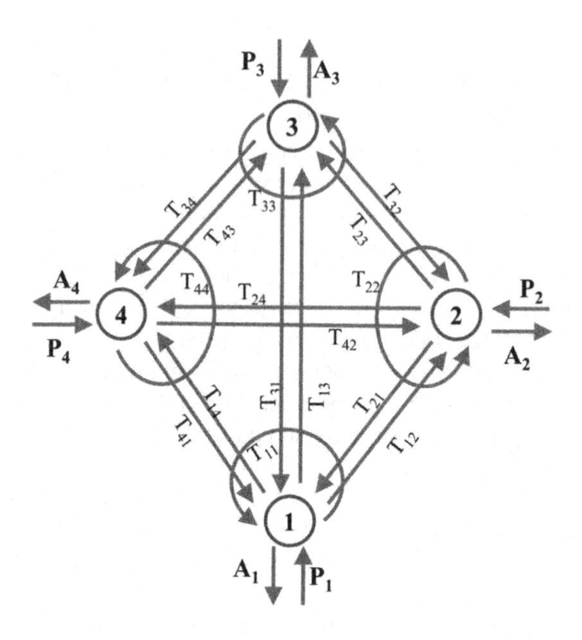

P_i: Traffic volume entering the intersection through node i
A_j: Traffic volume leaving the intersection through node j
$T_{i,j}$: Turning movement from node i to node j

Figure 10.2: Representation of signalized intersection turning movement counts as proposed in [45]

of absolute differences between observed and predicted traffic counts. The objective function is made to converge using a revised parameter optimization model. Then, they finally propose a genetic algorithm to impute turning movement counts.

The current state of research is lacking in several aspects. The datasets used are often small and don't represent a wide variety of traffic scenarios that may occur. Current methods use data aggregated at resolutions in the order of minutes and above and thus miss important details that may be found in more fine-grained data.

Our approach is different for two reasons:

1. Large number of samples collected using a realistic simulation framework
2. Results show that using more fine-grained data results in better accuracy for turning movement count prediction.

10.2 MACHINE LEARNING MODELS FOR TMC PREDICTION

As described in Chapter 6, we generate a dataset for turn movement counts to train neural network models. The same dataset is used here. The input to the models

Table 10.1

Table showing MSE on test set for prediction window of 375 seconds for different waveform aggregation levels.

Predicton Window	Waveform Aggregation level	MSE on Test Set
375 sec	5 sec	0.0002
375 sec	15 sec	0.0005
375 sec	25 sec	0.0002
375 sec	75 sec	0.0004
375 sec	375 sec	0.001

is detector waveforms (aggregated to some time interval) at stop bar and advance detectors. The input layer has 16 x (number of time steps) neurons, each neuron takes input from one detector and one time point.

The waveforms are represented as 1-D vectors, each with T components. Here T refers to the length of time a particular sensor's data is being considered, with each component being aggregated at a 5-second level.

We performed different experiments to analyze how the following parameters affect the TMC prediction. Tables 10.1 and 10.2 present these results. The first two columns in the tables represent the two features that are explained below.

- **Prediction window**: This defines the time window for which we predict the TMCs. Based on our results, we observed that as the size of the prediction window increases, the accuracy for predicting turning movement counts increases.

Table 10.2

Table showing MSE on test set for different prediction windows.

Prediction Window	Waveform Aggregation level	MSE on Test Set
375 sec	5 sec	0.0002
375 sec	15 sec	0.0005
750 sec	5 sec	0.0005
750 sec	15 sec	0.0006
750 sec	25 sec	0.0005
1000 sec	5 sec	0.0008
1000 sec	20 sec	0.0008

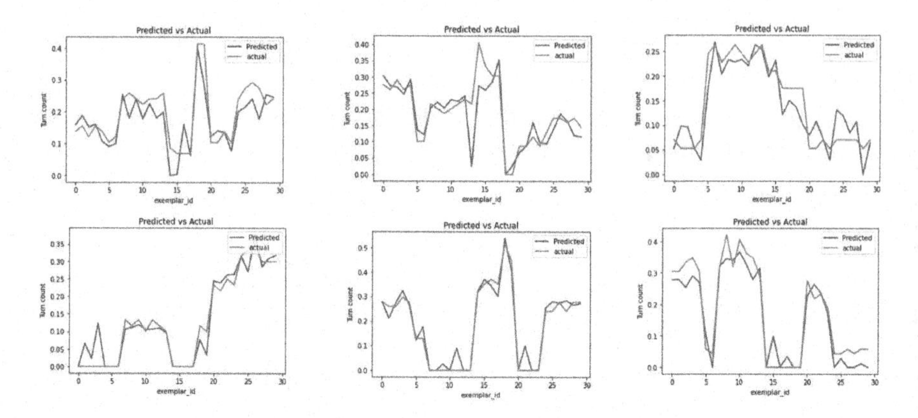

Figure 10.3: Actual vs. predicted turning movement counts for six different movements using the waveform model. The y-axis represents the turning ratio; the x-axis represents random exemplars of the test set. Each subplot corresponds to a particular direction/movement. Top row: East bound left, through and right. Bottom row: West bound left, through and right.

- **Waveform aggregation level**: This denotes the time resolution for each step of the waveform constructed from detector actuations. We vary this to find the optimal value of aggregation level for the waveform (5 sec, 15 sec, 25 sec, etc.).

We now describe a modular neural network architecture where input is transformed to the first hidden layer by sharing a common weight matrix among the detector waveforms of all the directions. The basic reasoning is that the detector waveforms (stop bar and advance) of the four directions would have similar properties. So, for the feed-forward, network we analyze the weights connecting the inputs to first hidden layer. Figure 10.8 shows the heat map of weights connecting the input layer to the first hidden layer of the feed-forward network. Figure 10.3 shows the actual vs predicted turn movement counts for different phases. In the plot the absolute value of weights is taken, multiplied by 100 to highlight the trends. The black marker in the plot separates the four directions, and the detector waveforms on the y-axis are ordered in correspondence to detectors in this order: east, west, north, and south. We can see from the plot that the weights connected to detectors on different directions look similar. Figure 10.4 showing percentage error for comparing different bins. This plot suggests that the model is performing well irrespective of the saturation level. Figure 10.5 shows MSE for eastbound through traffic for different bins, comparing different waveform aggregation levels. Figure 10.6 shows the percentage error for comparing different prediction windows. This plot suggests that accuracy increases when predicting for a larger window size.

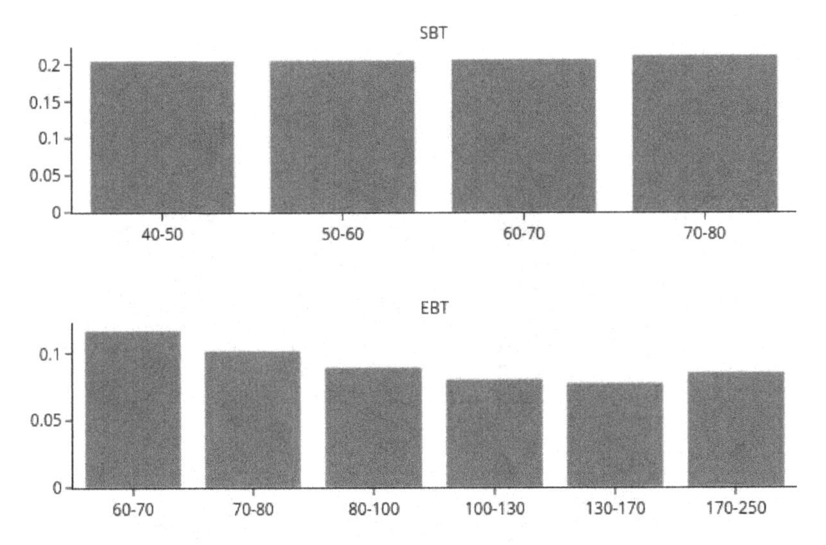

Figure 10.4: Plot showing percentage error for comparing different bins. This plot suggests that the model is performing well irrespective of the saturation level. SBT: South bound through; EBT: East bound through.

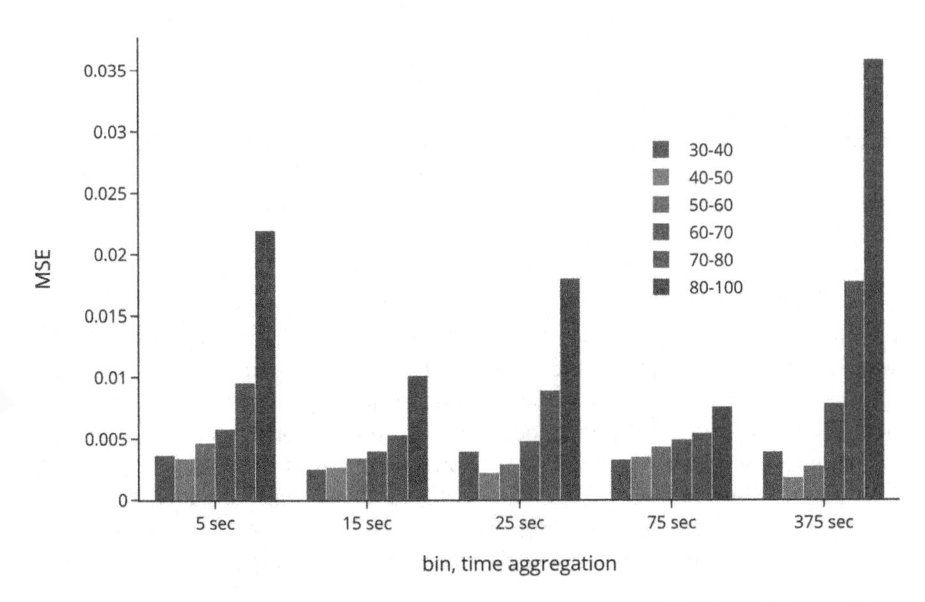

Figure 10.5: Plot showing MSE for eastbound through traffic for different bins, comparing different waveform aggregation levels.

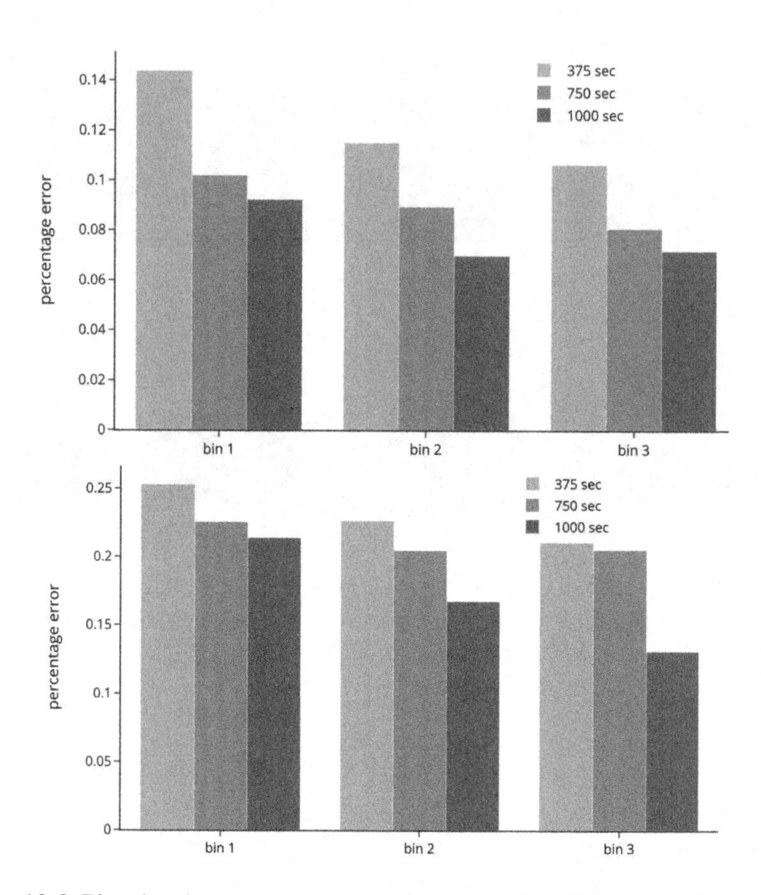

Figure 10.6: Plot showing percentage error for comparing different prediction windows. This plot suggests that accuracy increases when predicting for a larger window size. (a) EBT: East bound through; (b) SBT: South bound through.

So to decrease the number of learnable parameters and get better generalization, a common weight matrix is used for transforming detector waveforms on different directions to the first hidden layer. Figure 10.7 shows the comparison of percentage error for different bins comparing a fully connected network with the modular architecture. We can see that the modular architecture performs equally well with a lower number of parameters. Figure 10.8 shows heat map of weights connecting input layer and first hidden layer. The horizontal black bar separates the four directions. This indicates that the distribution of weights is similar for encoding waveforms of different directions and thus not much information is lost sharing the weights between all 4 directions.

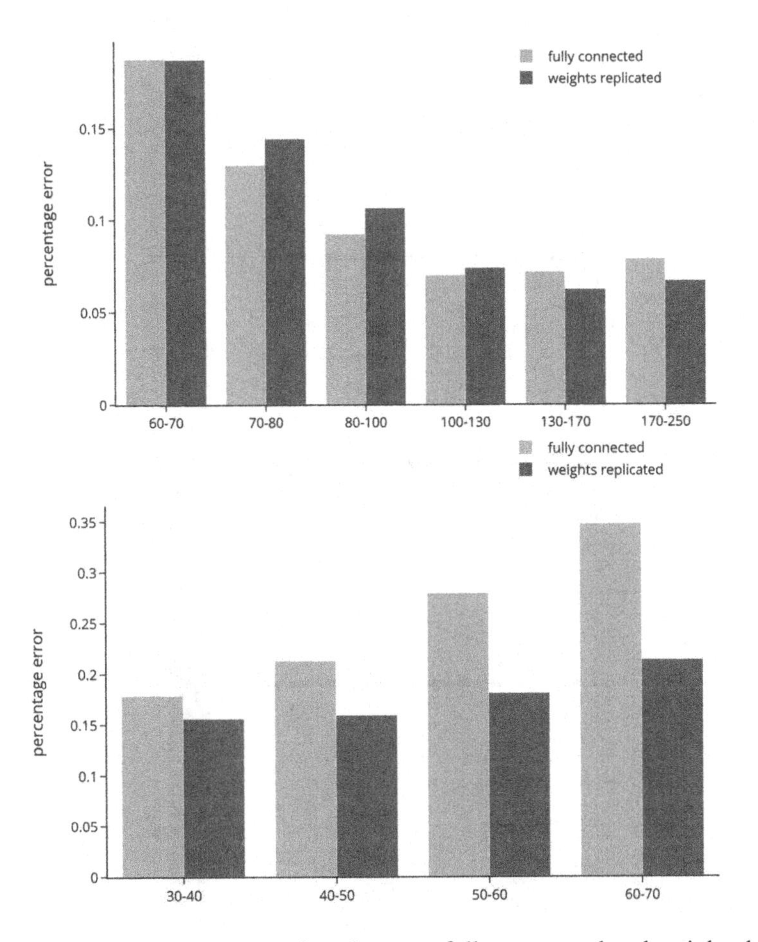

Figure 10.7: Plot showing comparison between fully connected and weight-sharing model. The weight-sharing model is comparable or better. (a) EBT: East bound through; (b) SBT: South bound through.

10.3 CONCLUSIONS

We have presented machine learning models for predicting turn movement counts for a single intersection, given information from stop bar and advance detectors. All of these data are available from ground sensors. Hence, this model can be easily adapted to practical datasets. The results show that using detector waveforms instead of aggregated traffic volumes leads to better accuracy for predicting turning movement counts. The neural network models developed perform well both under unsaturated and oversaturated conditions. Also, the use of a modular neural network architecture

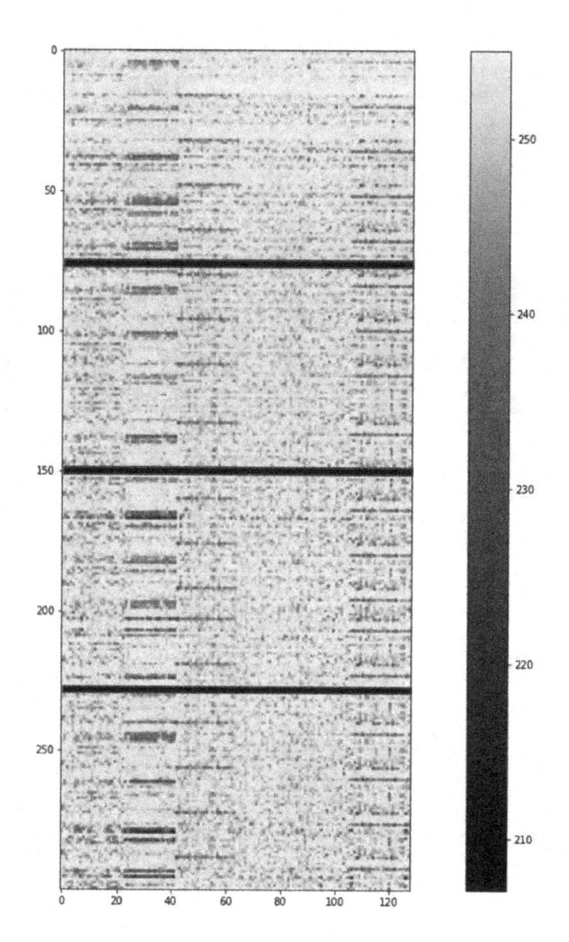

Figure 10.8: Plot showing heat map of weights connecting input layer and first hidden layer. The horizontal black bar separates the four directions. The ordering of detector waveforms along the y-axis is as follows: east, west, north, and south.

for predicting turning movement counts allows for generalization to new intersections. Instead of using different weight matrices across four directions (as in general feed-forward networks), we use a common weight matrix for all the four directions in the first hidden layer of the network. This modular architecture performs equally well and has fewer learnable parameters in comparison with a feed-forward network.

11 Coordinating Corridors

Abstract - The road network infrastructure (signal controllers and detectors) continuously generates data that can be transformed and used to evaluate the performance of signalized intersections. Current systems that focus on automatically converting the raw data into measures of effectiveness have proven extremely useful in alleviating intersection performance issues. However, these systems are not well suited for automatically generating recommendations or suggesting fixes as needed. In this chapter, we demonstrate the use of machine learning and data compression techniques to build a recommendation system. Specifically, we present an end to end solution for automatically generating intersection coordination plans.

Coordination involves synchronizing multiple intersections to enhance the operation of directional movements in a system [134]. Typically, the green times for a set of intersections are synchronized along the primary directions of movement using (speed-dependent) timing offsets to account for the travel time between intersections. Such coordination of intersections can result in improvements in the quality of traffic flow along an arterial or street [27]. The current practice for coordinating and re-timing intersections remains a largely manual process [27, 134] and is based on temporally limited data samples. In this chapter, we present a recommendation system that generates intersection coordination plans using high resolution intersection controller logs collected over large time periods.

The problem of coordinating intersections can be decomposed into two parts, namely:

1. Identification of intersections that should be coordinated and the times of day and days of week for which they should be coordinated. Coordination can have a negative impact on the quality of service on the side streets and hence should only be deployed when needed.
2. Offset computation: Once the set of intersections has been identified, green time offsets need to be computed such that the flow of traffic in the primary direction of movement can be maximized.

We describe an automated solution for the first problem, i.e., identification of intersections on a corridor that are good candidates for coordination and the time periods they should be coordinated for. Traditionally, to determine the intersections that should be coordinated and the time periods for which intersections should be coordinated, extensive data is collected for intersections, including human recorded observations. This can be both time and effort intensive, especially for large cities or regions with tens of corridors and thousands of intersections. Our approach, in this paper, is to leverage high resolution controller logs that are routinely collected at many intersections. These controller logs have been used to compute various measures of effectiveness for signalized intersections. Automated Traffic Signal Performance

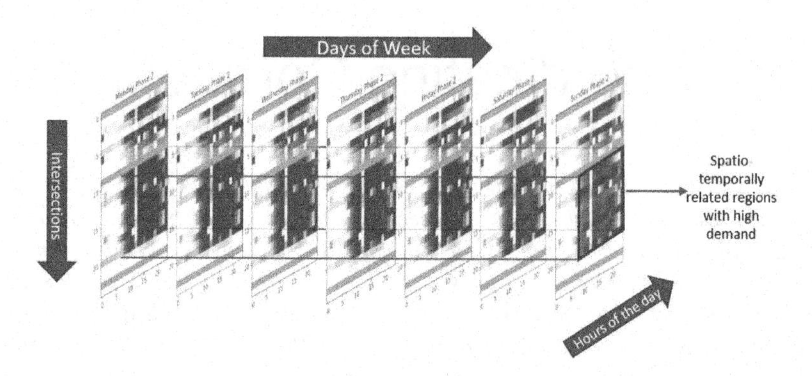

Figure 11.1: A visual representation of the three dimensional nature of the data. Each plane in this figure represents a different day of the week. Intersections are plotted on the y-axis and hours of day are on the x-axis. The pixel color represents the intensity of the demand and is visible in our ebook version. Cuboids, like the one highlighted in the figure, represent spatiotemporally related regions of nearly homogeneous high demand in these data. We are looking for such regions as they represent subset of contigous corridors that should be coordinated during particular days of the week and hours of the day.

Measures (ATSPM) systems (UDOT ATSPM [132] and its derivatives) also use these logs to automatically compute many intersection performance measures.

The problem of subdividing a single corridor into multiple spatially related sub-corridors with distinct traffic patterns or demand can be viewed as a 3-dimensional decomposition problem. The three dimensions in these data being time of the day (generally divided into hour size intervals or smaller), days of the week, and spatially correlated intersections. In essence, we are looking for spatially and temporally related regions with high demand in the 3-dimensional data shown in Figure 11.1. We need to look for similarity in traffic demand patterns across all three dimensions, while maintaining the spatial and temporal relations present in the data. This corresponds to cuboids with nearly homogeneous demand in Figure 11.1. And this will allow us to find sets of intersections (sub-corridors) that are good candidates for coordination because we expect them to have homogeneous performance. It will also allow us to deduce the times of day and days of week for which the coordination plans are required.

This approach has two major steps:

1. Building accurate, descriptive models for performance measures of interest using time series modeling for all intersections under consideration. This is done to ensure that conclusions are not based on temporally local observations because they may be susceptible to outliers.
2. Use of these descriptive models (based on key MoEs) to cluster the intersections in a corridor with similar traffic patterns (or demand). This is done for all corridors under consideration. We use the results of clustering and spatial

information within a corridor to automatically deduce contiguous regions in a corridor or sub-corridor where the intersections should be coordinated and the times of the day and days of week they should be coordinated for.

This approach is scalable and can be applied to thousands of intersections. And can reduce the manual effort required during the process of corridor coordination significantly. This should allow for frequent changes in coordination times as the traffic evolves over time.

The first step includes identification of key MoEs, and the computation of those measures from the controller logs. Next, we use time series modeling to build representative models for each intersection and MoE under consideration. This is detailed in Section 11.1. We use these models to then cluster intersections with similar traffic demand patterns. This is described in Section 11.2. The spatial and temporal decomposition for each corridor is also described in Section 11.2. Next, the results for a case study and a comparison with existing coordination plans is detailed in Section 11.3. Related work is presented in Section 11.4. The conclusion and future direction for this work are presented in Section 11.5.

11.1 INTERSECTION LEVEL MOE MODELING

We employ descriptive modeling of the traffic demand patterns and then use the models to build a compact 3-dimensional representation of these data. Descriptive time series modeling is highly suited for capturing the periodic variations in traffic demand patterns. This also reduces the susceptibility to outliers. Using historical data to model intersection demand over a period of time would give us representative vectors for each intersection (I) and day (D) by capturing all the trends and variations in traffic demand patterns. We first describe our approach followed by results achieved on data collected from real intersections.

11.1.1 DESCRIPTIVE MODELING

Based on input from practitioners and subject matter experts, we identified three key measures of effectiveness to model for this task. They are:

1. Demand-based split failures, computed using red and green occupancy ratios
2. Traffic volumes
3. Arrivals on red

These three MoEs were selected because they are useful in both coordinating intersections and diagnosing problems. Demand-based split failures and total volumes are both indicators of traffic demand patterns and hence these measures are modeled for each intersection. Modeling arrivals on red (number of arrivals when a phase is red) provides useful information to the practitioners, i.e. it is an indicator of correctness for the current (timing) offsets. Hence, we model all three measures. The first step is to aggregate the high resolution (10 Hz) data into minute-by-minute buckets. For split failures reported on a phase, a value of 1 is recorded during the minute the failure was reported. This is also true for minutes when more that one (demand based) split failure is reported. A 0 is recorded if there are no reported split failures for the phase. Using the computation in the previous paragraph (ROR/GOR), the split failures reported in coordinated corridors (when there is no demand) are ignored. Hence, we compute the duration of split failures in minutes and store that information as a time series vector. Similarly, we compute the traffic volumes and arrivals on red. The output of the first step is vectors with 24×60 (1,440) entries representing the behavior of a particular phase over the entire day. The next step in our methodology is to take these $1,440$-digit feature vectors and aggregate into one-hour bins. The dimensionality of the new vectors is 24, i.e., each vector has 24 entries representing one intersection for each the hour of the day. Note that the one-hour bin size was selected based on input from subject matter experts. However, various other bin sizes (15 min or 30 min) can also be explored. In our analysis, we have considered only the primary directions (phases 2 and 6) while creating these vectors and concatenated the vectors for phase 2 and phase 6 to create one vector for each MoE of interest.

The modeling step is critical because these models are effective in capturing time of the day and day of week variation in traffic demand patterns without being susceptible to temporal outliers. Hence, the modeling improves the current practice, which is based on temporally limited observations (one to two weeks of data).

We now describe the modeling step in more detail. The data that we aim to model are periodic in nature. Based on the previous research and input from subject matter experts, there are two periodic trends that need to be modeled: the daily and weekly periodicity of the data. Hence, we deploy Fourier series to model our data. The model can be described as follows:

$$y(t) = g(t) + s_1(t) + s_2(t) + e(t)$$

where,

$y(t)$ is the target variable, MoE in our case
$g(t)$ is the term that models growth over time
$s_1(t)$ is the term that models daily periodicity
$s_2(t)$ is the term that models weekly periodicity
$e(t)$ is the error term.

The periodic terms $s_1(t)$ and $s_2(t)$ can be further expanded as:

$$s_i(t) = \sum_{n=1}^{N}(a_n cos(\frac{2\pi nt}{P}) + (b_n sin(\frac{2\pi nt}{P}))$$

This modeling approach is adaptive to different periodic effects because of the reliance on Fourier series. The parameter P controls the periodicity of the time series under consideration. For the first periodic term, $s_1(t)$, the value of P is set to 24 to allow it to model daily trends. And for the second periodic term, $s_2(t)$, the value of P is set to 7 to allow it to model weekly trends. We build descriptive models for each intersection and measure separately.

Prophet™, an open source, state of the art, time series modeling and forecasting tool that was used to train these descriptive models. Periodic time series modeling in Prophet is also based on Fourier analysis, and hence it is well suited for our needs. To ensure the quality of the models, a test-train split was created in the dataset. Specifically, we used eight weeks of data to train the models and tested the models using the next four weeks of data. In all cases, the model accuracy is above 90%. This gives us a high degree of confidence that the models are capturing the periodic trends in the data. This approach allows us to build representative models for each intersection and MoE combination, thus enabling the comparisons across intersections that will be useful in the next steps.

11.1.2 EXPERIMENTAL RESULTS

Figure 11.2 is a visualization of the daily and weekly trends for all three measures of effectiveness that were selected. We expect that proximal intersections will report similar trends for split failures and volumes unless they are saturated. We can also observe that the volumes may be higher for certain days of the week, but the intersections are saturated (split failures) during the same times for all days of the week.

In the next section, we use the descriptive models built for demand-based split failures to cluster together intersections on the same corridor that are busy during the same times of day and days of week and then derive spatial decomposition of the corridors using this information.

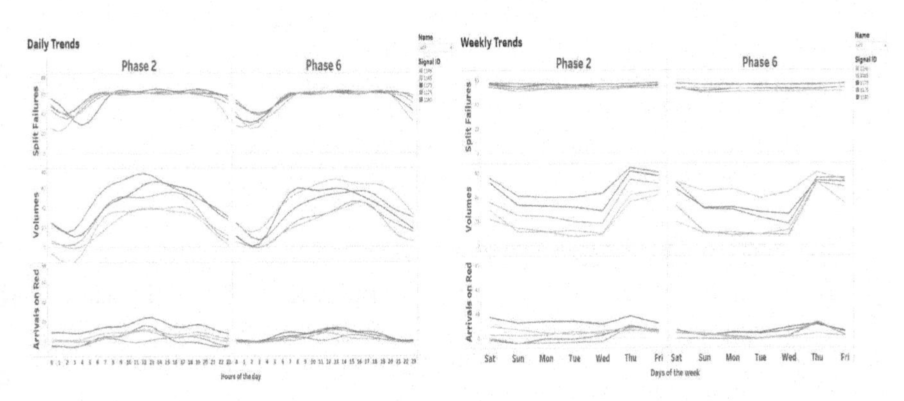

Figure 11.2: A visualization of the daily and weekly trends captured by the models. Based on the daily trends for split failures and volumes, we can see that this set of intersections is busy from 5AM to end of the day. Arrivals on red are not correlated to the other two measures as they depend on the value of the timing offset. Although, not the focus of this paper, this can be useful in deriving the duration for each intersection where the offset can be improved. We can also see that most intersections report extended periods of saturation on all days.

11.2 CLUSTERING AND SPATIOTEMPORAL CORRIDOR DECOMPOSITION

This section details the clustering of the intersections on the same corridor for different hours of day and days of the week using demand-based split failures. We pick this measure because it is a very good indicator of duration during which a intersection is unable to serve all the traffic demand. It is worth noting that other measures of traffic congestion can be used instead and the rest of the description does not depend on this choice.

Let there be multiple intersections $I = I_1, I_2, I_3, \ldots, I_n$, each of which is represented by its location (geographical coordinates). Let $M\ [I,\ D,\ H]$, D ∈ [1,7] and H ∈ [1,24] represent the modeled demand based split failures for a set of intersections for different days of the week and hours of the day. Although, we use hours as the discretization for decomposing the time of the day, one can use other interval sizes (15 minutes, 30 minutes) instead. These data can then be represented as a three dimensional cube, the three dimensions being intersections, days of the week and hours of the day.

For a set of intersections to be coordinated, they must be spatially proximal. However, the problem of finding spatially proximal regions of high demand in this 3D cube can be challenging. Hence, we first perform time of day clustering for all intersection-day (I-D) pairs. As a result of this, we obtain a single cluster identifier which represents the intersection demand patterns for the whole day. By clustering the I,D pairs, across the hour of day dimension we obtain a representative (cluster ID) for each pair and let this matrix be C[I,D].

Next, we deploy day of week (DoW) clustering to deduce which days of the week report similar demand patterns, and this effectively reduces our problem to a single dimension. This corresponds to looking at similarity between columns of C[I,D].

For each subset of columns found in the previous step, we look for nearby rows of the C matrix with similar behavior to derive subsets of similar intersections across the corridor with nearly homogeneous demand to generate candidates for coordination. A detailed explanation of each of these steps is now presented.

11.2.1 HOUR OF DAY CLUSTERING

As described above, we use the vector $M\ [i, d, *]$ to represent the performance of an intersection (i) for a single day (d). We propose the following steps to cluster the intersection-day (I-D) vectors and reduce our problem from a 3-dimensional to a 2-dimensional space:

1. First, we discretize the performance of each intersection, for each day, using demand-based split failures. Specifically, any observation that is $x\%$ of the peak demand for that intersection-day combination is classified as busy. We tried many values of x before settling on 55%. In this context, the peak demand is defined as mean of the three busiest hours under observation. These discrete vectors are represented as $B\ [I, D, H]$.
2. The next step is to deploy top-down hierarchical clustering combined with a custom distance measure to group similar intersection-day vectors together.

3. The last step is to compute the time of day cluster representative, $R[H]$, for each cluster. We compute the first eigenvector of set of intersection (I), day (D) vectors that get clustered together. This cluster representative, $R[H]$, summarizes the time of day traffic demand patterns for each cluster.

We use normalized Hamming Distance between vectors $B\ [i_1, d_1, *]$ for $B\ [i_2, d_2, *]$ for the purposes of clustering. This represents the number of hours for which the traffic congestion is different for two intersections. Hamming distance between two discrete vectors is defined as the number of positions the two arrays are different.

This measure enables us to compare the times of day for which the two intersections are performing similarly. For example, if $B\ [i_1, d_1, *] = [1, 1, 0, 0, 0, 0, 1, 1]$ and $B\ [i_2, d_2, *] = [1, 1, 1, 1, 0, 0, 1, 1]$, the normalized distance will be $\frac{4}{8} = 0.5$. Figure 11.3 showcases how the clustering reduces the 3-dimensional problem to 2 dimensions.

Another important outcome of clustering is a representative vector for each cluster Y represented by vector $R[Y, H]$. $R[Y, H]$ corresponds to the hours when any intersection, day that is part of cluster Y is congested (and potentially has to be coordinated).

Next, we present the results of clustering for a sample corridor. $C\ [I,\ D]$, represent the time of day clustering results for all the intersections on a single corridor. Figure 11.3, part (B), is a pictorial representation of the matrix $C\ [I,\ D]$, i.e., the clusters that were discovered on a sample corridor. Each row of the image represents an intersection for 7 days. And the pixels are colored according to the respective cluster ID. The cluster ID's are ordered based on the increasing order of traffic demand. The rows that are grayed out represent bad data. We can see that two major regions of distinct intersection performance exist in this corridor. These are highlighted as S1 and S2.

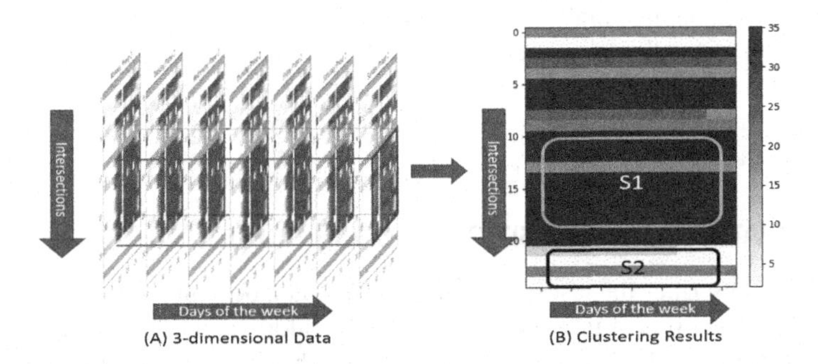

Figure 11.3: Reduced representation of three dimensional data (A) into two dimensions. Additionally, (B) also presents the clustering results for a single corridor. Each row represents an intersection in the corridor and the columns represent the days of the week. Each cell is colored according to its cluster ID, with the colors visible in our ebook version. The gray color represents intersections in the corridor for which data is not available. S1 and S2 represent two regions of distinct intersection performance in this corridor. Our goal is to find these regions automatically.

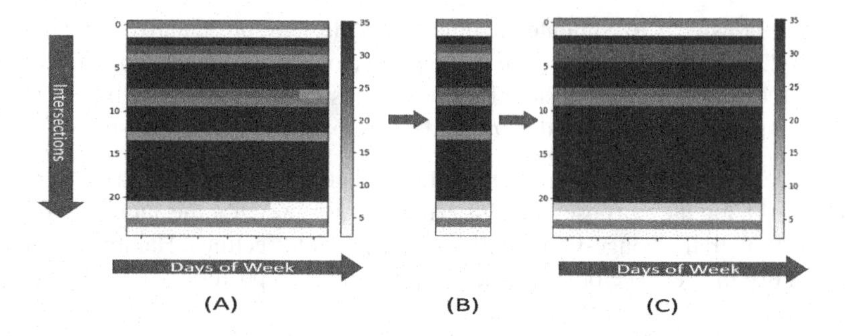

Figure 11.4: The results of the algorithm used to detect spatially proximal, nearly homogeneous regions within a corridor such that these intersections can be coordinated together. (A) represents the results of the time of day clustering, (B) represents the results of the day of week clustering, and (C) depicts the regions detected by the spatial decomposition algorithm.

11.2.2 DAY OF WEEK CLUSTERING

Next, we use the results from the previous section and look for (not necessarily contiguous) days that are behaving similarly. Recall that $C\ [I,\ D]$ represents the time of day clustering results for all the intersections on a single corridor. This corresponds to the matrix presented in Figure 11.3. We cluster the seven, $C[I,*]$ vectors to find days of the week that report similar demand patterns.

The distance between $C[I,\ d1]$ and $C[I,\ d2]$ can then be defined by the number of dissimilar elements in the two vectors. However, a much better measure is to use the corresponding representative vectors for the two cluster ids in this computation. Since each elements corresponds to a cluster id, this distance for two related cells is computed based on hamming distance between $R[C[I,\ d1],*]$ and $R[C[I,\ d2],*]$.

We can then use this measure to cluster all seven $C[I,*]$ vectors. This gives us the days of the week with similar demand patterns. Figure 11.4, parts (A) and (B), present a visualization of this process. For this example, we find only one cluster across all 7 days.

11.2.3 SPATIAL DECOMPOSITION OF CORRIDORS

This section describes the last step of this process, i.e., decomposing a corridor into spatially proximal sets of intersections (or sub-corridors) which are good candidates for coordination. Each corridor has constraints on the number of coordination plans that can be implemented on the corridor simultaneously. Too many sub-corridors have negative impact on the quality of progression of vehicles. Hence, we look for spatially proximal intersections with nearly homogeneous performance. This is done separately for each cluster of days founds in the previous step. The clustering is achieved by clustering adjacent rows of $C[I,D]$ (for only a subset of days) with similar behavior. Again, the distance between is computed using the hamming distance between corresponding R vectors as in the previous step. In addition to the above approach, the clustering approach allows for a small number of non-conforming intersections within a cluster of similarly behaving intersections. This is achieved by a two phase clustering where only similar adjacent intersections are clustered together. This is followed by merging two clusters with similar behavior if they are separated by a single intersection whose traffic congestion behavior is lower than the two adjacent clusters on either side that should be merged. This process is depicted by Figure 11.4, part (B) and part (C).

This concludes the automated approach for deducing coordination plans for a set of intersections on the same corridor and the times of day and days of week for which they should be coordinated. The next section presents a case study which showcases the results of this approach.

11.3 RESULTS AND CASE STUDIES

The techniques described in the previous sections were applied to approximately 200 intersections spread across 10 key corridors in Orlando, Florida. This section presents a case study for Lake Mary Boulevard which showcases our results. For the year 2019, we compare our results with coordination plan used during that year.

The results presented in Figures 11.5 and 11.6 were based on descriptive models built using data from 2019. Figure 11.5 details the clustering and spatial

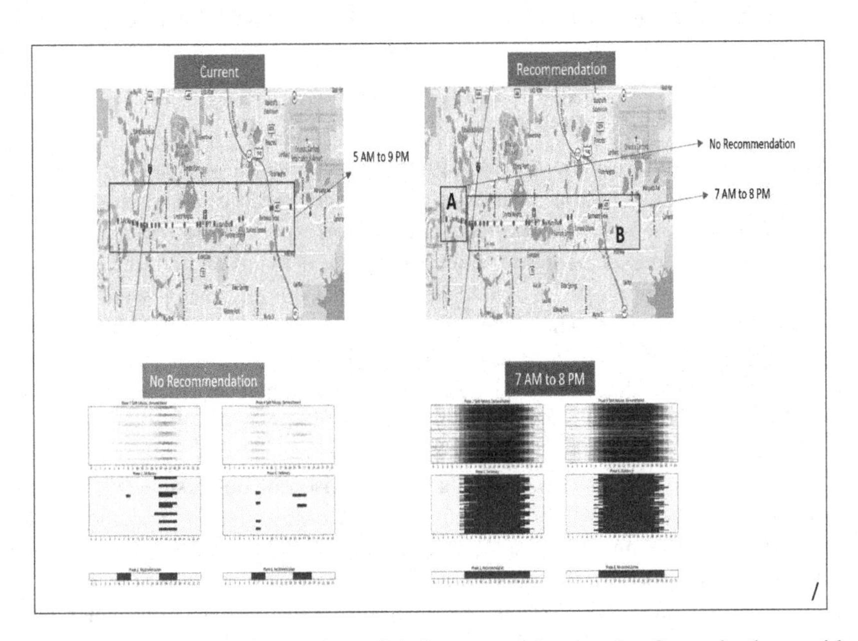

Figure 11.5: The clustering and spatial decomposition results for a single corridor (Lake Mary Boulevard) using data from the year 2019. The two major sub-corridors are highlighted on the map. The split failures for the major (2,6) and minor (4,8) phases have also been plotted on the left. It should be noted that our clustering is based on the major phase split failures only.

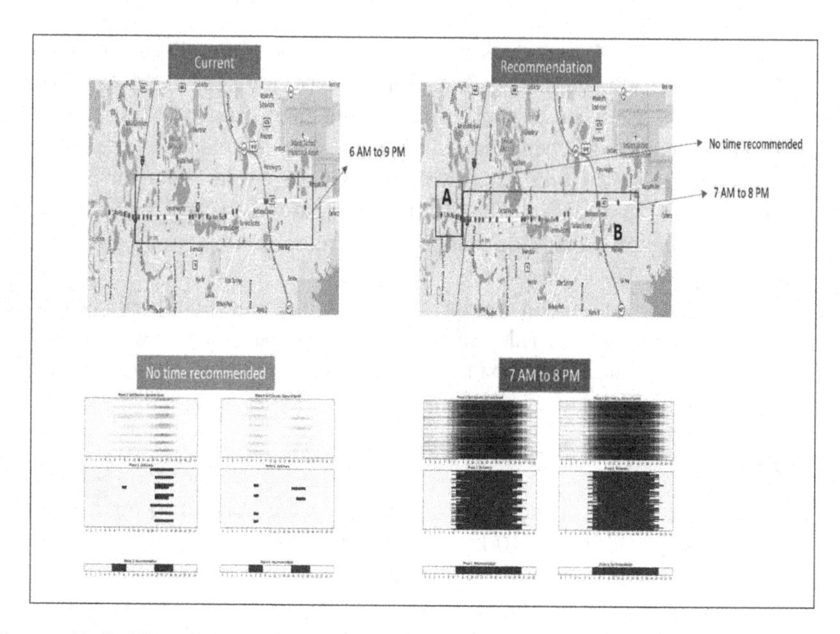

Figure 11.6: We present the recommended coordination plan based on the data collected for Lake Mary Boulevard for the year 2019. Intersections in sub-corridor (B) are currently coordinated between 6 AM and 9 PM. Whereas the intersections in sub-corridor (A) are not coordinated for most times of the day. Thus, our recommendation can improve the existing coordination times.

decomposition results for the corridor under consideration. We can see that one major (and one minor) region of distinct performance are present on this corridor. These are marked as A and B. The intersections corresponding to these regions are highlighted on the map. This figure also presents the demand on both the major and minor phases for comparison. It should be noted that our clustering is based on the major phase demand only. The minor phase demand can be useful in further sub-dividing these regions and hence is presented with our results. These results show that for subcorridor represented by B, the signals should be coordinated from 7 AM to 8 PM (this is shorter than the currently used times of 6 AM to 9 PM). For A, no coordination is necessary.

The next figure, Figure 11.6, presents a comparison of our results with current coordination plans can be improved using our recommendation. Specifically, the coordination time for one of the sub-corridors can be reduced. Choosing appropriate periods for coordinating signals on a corridor can have positive impact on the quality of service on the side streets. We also present the clusters from which these coordination plans were derived.

11.4 RELATED WORK

Traffic engineers often face the challenge of quantifying the performance of signalized intersections and effectively troubleshooting day-to-day operational issues. In many cases, problems that are particularly difficult to troubleshoot are the ones that reoccur only at specific time of day or because of exclusive traffic patterns. A practitioner has to generate and analyze performance measures for each direction of movement, and for every intersection manually to understand and troubleshoot thee problems. There have been past attempts to address these shortcomings. Specifically, data analytics techniques have been previously applied to traffic flows and we now present a few relevant studies. Howell Li et al. [80] present a heuristic based on system wide split failure identification and evaluation. By using this heuristic, they demonstrated performance improvements for specific corridors. The paper titled Graphical Performance Measures for Practitioners to Triage Split Failure Trouble Calls [41], demonstrates the use of graphical performance measures based on detector occupancy ratios to verify reported split failures and other shortcomings in signal timing that are often reported to traffic engineers by the public. Wemegah *et al.* [147] present techniques for management of large signalized intersection datasets with the aim of analyzing traffic volumes and congestion, addressing all the steps in the analytics pipeline, namely, data acquisition, data storage, data cleaning, data analysis and visualization. Huang *et al.* [56] propose a set of new, derived MoE's that are designed to measure health, demand and control problems on signalized intersections. The newly proposed MoE's are based on approach volume and platooning data derived from ATSPM [132] systems.

11.5 CONCLUSION

In this chapter, we presented an automated framework for deducing intersection coordination plans. To this end, we leveraged time series modeling to build descriptive models for all the performance measures under consideration. Next, these models are used to cluster intersections with similar traffic demand patterns. Furthermore, we spatially decomposed the corridors into regions with near homogeneous traffic demand, generating sets of intersections that are good candidates for coordination. Lastly, we presented a case study to showcase our recommendations and compared these recommendations with the currently deployed coordination plans. This automated approach can be deployed in two ways. First, it can serve as a decision support system for traffic engineers and generate sets of intersections that are good candidates for coordination. Second, it can serve as a prior step for an automated signal re-timing solution.

12 Modeling Input Output behavior of Intersection

Abstract - Traffic flow dynamics in the vicinity of urban arterial intersections is a complex and nonlinear phenomenon, influenced by factors such as signal timing plan, road geometry, driver behaviors, etc. Predicting such flow dynamics is an important task for urban traffic signal control and planning. Current methods use microscopic simulation for studying the impact of a large number of signal timing plans at each of the intersections. A major drawback of microscopic simulation is that they are they are based on source destination traffic generation models and cannot incorporate the high resolution loop detector data such as that are provided by automated traffic signal performance measures (ATSPM) based systems. The arrival (or departure) information of each vehicle on a detector can be thought of as a time series waveform. Given the high granularity of ATSPM data, this waveform can be used to several interesting analyses. The waveforms can be used to derive information on platoon dispersion as vehicles progress across the corridor. Also, these waveforms can be modelled to understand how the vehicles progress across the corridor for a variety of signal timing plans.

In this chapter, we present neural networks models that can can be used to effectively leverage the waveforms collected at multiple sensors (stopbar and advanced) on the intersection to model the traffic dynamics both at an intersection and across intersections. We describe how modelling of these waveforms can be useful to understand traffic flow dynamics under different signal timing plans and can be potentially integrated into signal timing optimization software. Further, these methods are three to four orders of magnitudes faster than using microscopic simulations.

Reducing congestion requires coordinating the signal timing plans of each intersection on a corridor or network so that most of the vehicles do not have to wait at traffic intersections (effectively by changing the cycle length, splits, and offsets). This is a complex problem as changing the signal timing on one intersection affects the traffic on the adjacent intersections. Thus, it requires choosing the appropriate combination of signal timing plans for each intersection on a corridor or a network. Microscopic simulations are useful in understanding the correlated impact of signal timing plan at multiple intersections on the overall performance of traffic movement. They are able to model the output traffic patterns (along all outbound lanes) of an intersection based on the traffic patterns of the input lanes. Additionally, they can capture the dispersion of platoons when the vehicles move from one intersection to another. Both of these are important requirements for corridor and network optimization.

A major drawback of microscopic simulation is that they are they are based on source destination traffic generation models and cannot incorporate the high-resolution loop detector data such as that are provided by automated traffic signal

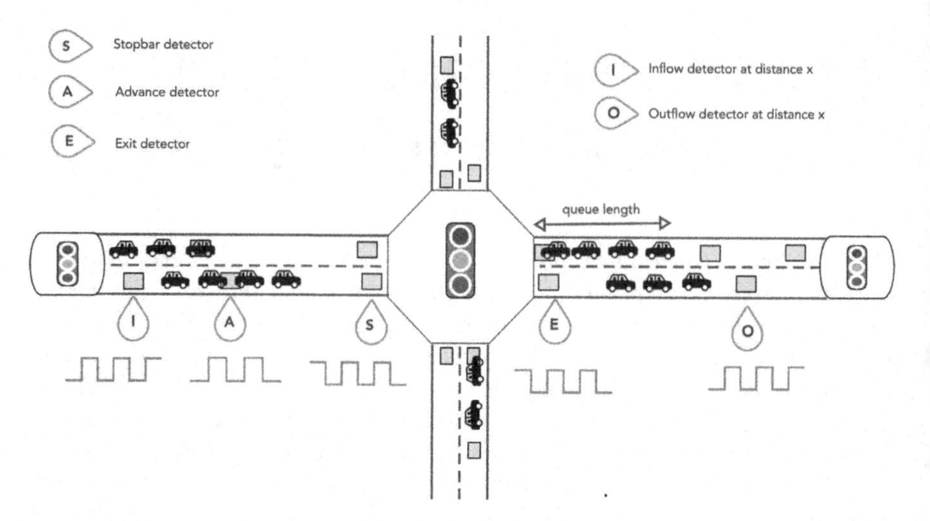

Figure 12.1: Generic single-lane intersection. In real world, the vehicle waveforms is generally observed at stop bar, advance detectors. We also introduce a set of virtual detectors - exit, inflow, outflow detectors. These are the locations where the waveforms are imputed so that we can model the traffic flow dynamics between any pair of intersections.

performance measures (ATSPM) based systems. Unlike origin-destination models these systems can provide traffic arrivals and departures at stop bar detectors (and in many cases advanced detectors) at 10 Hz (and are effectively much more precise than the origin-destination models in capturing the traffic variations throughout the day). Although some simulators allow this for a single intersection, they cannot achieve this for corridor and network simulation where neighboring intersections have input-output relationships. Microscopic simulation, by their nature, are sequential in a nature and the time requirements are proportional to the number of vehicles in the system and are relatively slow.

The arrival (or departure) information of each vehicle on a detector can be thought of as a time series waveform (see Figure 12.1). The waveform at the exit of the first intersection depends on the arrival waveforms in all the directions along with signal timing state at the intersection (Figure 12.2. As this waveform progresses to downstream intersection it is further affected by platoon dispersion, addition of more vehicles and signal timing state at the downstream intersection. Given the high granularity of ATSPM data, this waveform can be used to derive information about platoons (multiple vehicles passing without significant distance) or gaps (no vehicles passing through for a duration). In this chapter, we describe how deep neural networks based machine learning systems can be used to effectively leverage the waveforms collected at multiple sensors (stop bar and advanced) on the intersection to model the traffic dynamics both at an intersection and across intersections. In particular, using these waveforms at stop bar and advance detectors:

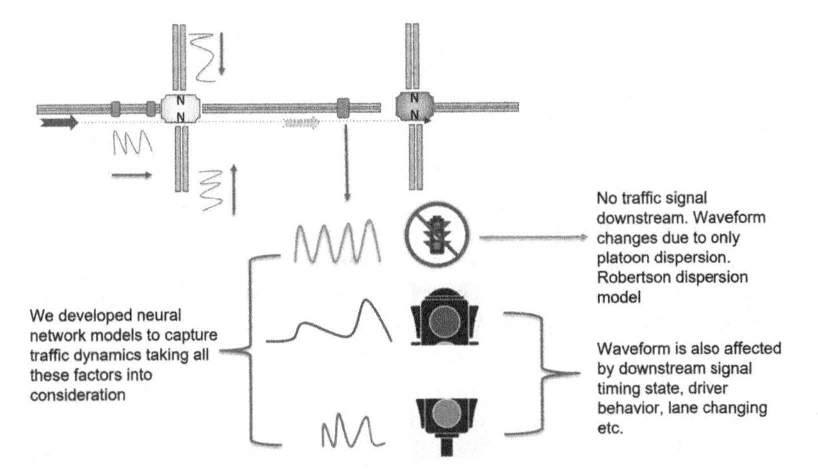

Figure 12.2: The arrival (or departure) information of each vehicle on a detector can be thought of as a time series waveform. The waveform at the exit of the first intersection depends on the arrival waveforms in all the directions along with signal timing state at the intersection. As this waveform progresses to downstream intersection it is further affected by platoon dispersion, addition of more vehicles and signal timing state at the downstream intersection.

- We present models to both impute the traffic waveform from each inbound direction to an intersection as well as the traffic waveform to each outbound direction from the intersection conditional on the signal timing plan. The input waveforms in all the directions can be used to understand if the current signal timing is near optimal. The models can be used to model the vehicle progression to downstream intersections and estimate performance measures for different signal timing plans. However, since they use data that is directly measured based on the traffic sensors in the network, they are a much more accurate indicator of traffic movement than imputed origin destination pairs.
- We present models that can predict the dispersion along a road segment (exit from one intersection to entrance of the neighboring intersections) more accurately than the Robertson model [114] that is traditionally used. This is because our models, like microscopic simulation models, can effectively capture non-uniform velocities of vehicles as well variation in driver behaviors.
- We present models that can predict the impact of signal timing of the upstream intersection on platoons as they arrive close to a downstream intersection. The output waveform from a given intersection along with the output waveform on the downstream also be used to understand the leakage or addition of traffic during a short time period.

These models can capture both the local (i.e., near an intersection) traffic flow dynamics as well as coupled traffic flow dynamics (i.e., between two consecutive intersections) and are significant extensions of the prior work on platoon dispersion models. We provide multi-scale error measures to demonstrate that our predictions are accurate and comparable to microscopic simulation.

These models use novel deep learning based architecture with attention layers [16] and teacher forcing [149] that is specifically designed to model and predict the behavior of input and output behavior at an intersection using advance and stop bar high resolution loop detector data and signal timing information. The use of our GPU implementation of deep neural networks can generate accurate predictions at three to four orders of magnitude faster than using microscopic simulations for this purpose.

These models leverage both the recent advent of automated traffic signal performance measures (ATSPM) [76] systems provide this information and signal timing information at a decisecond level and the availability of GPU-based computing and deep neural network algorithms to model traffic behavior.

Adaptive traffic signal control software's used in practice (TRANSYT-7F, SCOOT etc.,) leverages Robertson platoon dispersion models to predict vehicle arrival rate at downstream intersection and use that to calculate optimal signal timing parameters like cycle length, offset, green splits etc. The effectiveness of these tools depends on platoon dispersion models in terms of how well they predict the progression of vehicles downstream, which our models can provide better accuracy. Thus, our methods can be integrated into other signal timing optimization frameworks, such as those based on linear models and reinforcement learning for corridor and network coordination optimization.

12.1 RELATED WORK

Machine learning techniques, including deep neural networks, have been successfully applied to traffic data for traffic state prediction [142] for short-term (5-30 minutes), medium-term (30-60 minutes) and long-term (1+ hour) time windows.

Most of the previous work is in volume prediction or predicting flows at downstream intersections. We outline this work below:

1. Predicting volumes at cycle level: A number of techniques have been used for predicting volumes at cycle-length resolution (generally 2 minutes). This includes ensembled kernelized matrix completion [81], shockwave analysis, and Bayesian networks [139].
2. Predicting volumes at 5-minute intervals: Techniques that have been used include ARIMA [141], deep learning with non-parametric regression [3], multisegments (with recurrent and convolutional layers), deep neural network [140], graph embedding coupled with a generative adversarial network [155], and a combination of linear genetic programming (LGP), multilayer perceptron (MLP), and fuzzy logic [161].

However, these works primarily focus on predicting volumes for the same loop detector or location at a future point in time. They do not attempt to model the outflow waveform exiting a signalized intersection or its modification downstream.

Somewhat relevant to our work are Ehlers [34] and Wright et al. [151], which use geometric deep learning architectures, and Sun and Zhang [125], which uses a linear model to predict flow at the downstream intersection, given upstream intersection detector flows.

Platoon dispersion models have been proposed in the literature to model flow rates of a platoon as it traverses through a corridor. Lightlhill and Witham modelled platoon dispersion using kinematic wave theory [83]. Platoon dispersion models as analogous to continum fliud based on shock wave theory is proposed by Pacey [107]. The model that is being widely used is based on Robertson platoon dispersion model [114]. Some recent studies also proposed variations of Robertson's model to account for heterogeneous traffic flow condition [61, 153, 154]. These dispersion models however are not targeted towards determining the impact of signal timing or traffic entering from the side streets. The flow rates at a point are similar to waveforms described above. By using a composition of deep neural networks and Robertson model, our novel approach can incorporate the impact of signal timing of the next intersection on the platoon dispersion (cf. Section 12.2).

12.2 PROPOSED MODELS

Vehicle loop detectors that have traditionally been deployed at intersections to detect the passage of vehicles can measure the absence or presence of vehicles passing above them. The arrival (or departure) information of each vehicle on a detector can be thought of as a time series waveform (see Figure 12.1). Table 12.1 describes notations of different variables used in the modelling and Table 12.2 describes the input output combinations for different proposed models. This waveform can be used to provide information about platoons (multiple vehicles passing without significant distance) or gaps (no vehicles passing through for a duration. These waveforms are only available at advance detectors and stop bar detectors, as there are typically no detectors available at inputs and outputs (some U.S. states have these detectors, but most do not).

To model the progression of vehicles between intersections, in terms of input and output waveforms, we introduce a set of virtual detectors placed at an intersection (Figure 12.1) - Exit detector, Inflow detector, Outflow detector. The underlying idea is that we train the neural network models based on simulated data to be able to impute waveforms at these virtual detectors using data observed at stop bar and advance detectors. Imputing waveforms at these virtual detectors helps us model the traffic flow dynamics between a pair of intersections independent of distance between them and also understand the progression of vehicles for a variety of signal timing plans (the waveforms observed at stop bar and advance detectors is correlated with signal timing). Our focus is on making predictions of these waveforms at the granularity of approximately 5-10 seconds (generally, this corresponds to a range of 0-3 vehicles). Our experimental results show that we can achieve this with a high level of accuracy.

In this work we propose non-parametric neural networks to model the progression of vehicles between signalized intersections. We decompose the modelling of platoon dispersion between intersections using 4 different models (Figure 12.3) based on

Table 12.1
Table describing notations of different variables used in the modelling.

Name	Description	Agg. level	Dimension	Type
S	Waveform at stopbar detector	5 sec	1×150	Integer (0-5)
A	Waveform at advance detector	5 sec	1×150	Integer (0-5)
SIG	Signal timing information	5 sec	8×150	Binary (0,1)
TMC	Turning movement counts ratio	750 sec	1×12	Integer
INF_d	Inflow waveform at distance *d* upstream the intersection	5 sec	1×150	Integer (0-5)
EXIT	Waveform at virtual exit detector	5 sec	1×150	Integer (0-5)
OUT_d	Outflow waveform at distance *d* downstream the intersection	5 sec	1×150	Integer (0-5)

where the observations and collected and where predictions are made. Description of each these models is presented in the following subsections. Figure 12.3 shows input-output relationships for different models. Given waveforms at stop bar and advance detectors and signal timing information, the models should be able to predict waveforms at the exit of the intersection and the waveform at a certain distance downstream (outflow) and to reconstruct inflow at a certain distance upstream. The description for each of the models follows.

12.2.1 STOP-BAR-TO-EXIT WAVEFORM PREDICTION MODEL

Stop-Bar-to-Exit Waveform Prediction Model (M_{exit}), this model predicts the waveform that exits the intersection, given (1) the waveforms at stop bar detectors (S)

Table 12.2
Table describing input and output variables for different models as shown in Figure 12.3

Name	Description	Inputs	Outputs
M_{exit}	To predict waveform at exit detector	S, A, SIG, TMC	OUT_d
M_{in}	To reconstruct inflow waveform	S, A, SIG	INF_d
M_{down}	To predict progression of exit waveform towards downstream intersection	OUT_0, d	OUT_d
M_{sa}	To impute waveform at stopbar, advance detectors	INF_d, SIG	S, A

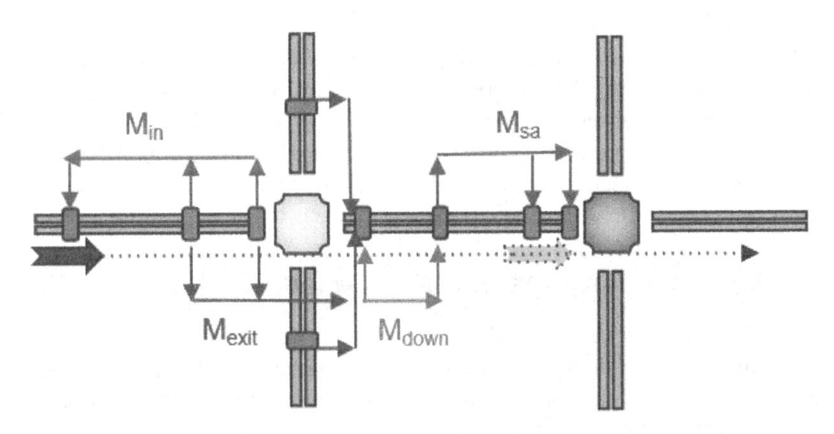

Figure 12.3: Diagram showing input, output relationship's for different models as proposed in Section 12.2. We decompose the modelling to capture both the local (i.e., near an intersection) traffic flow dynamics as well as coupled traffic flow dynamics (i.e., between two consecutive intersections). M_{exit} - To model the waveform at exit detectors. M_{in} - To reconstruct inflow waveform. M_{down} - To predict progression of exit waveform towards downstream intersection. M_{sa} - To impute waveform at stopbar, advance detectors.

and advance detectors (A) along all the directions, (2) the signal timing plan of the intersection and (3) turning movement counts. This model captures the local effects at the intersection. In practice, intersections do not always have an exit detector, but nevertheless, we instantiate one in our simulator in order to gain an in-depth understanding of flow dynamics local to an intersection.

12.2.2 INFLOW WAVEFORM RECONSTRUCTION MODEL

Inflow Waveform Reconstruction Model (M_{in}), this model reconstructs the unperturbed inflow waveform at an intersection, given the observed waveforms at the intersection's stop bar and advance detectors and the signal timing plan. The inflow waveform can be thought of as the incoming waveform that has exited the upstream intersection and is still sufficiently far away from the intersection of interest and thus is not yet affected by the queues and signal timing plan of the intersection of interest.

An important application for reconstructing the inflow is to potentially use it for signal timing optimization. The observed stop bar and advance detector waveforms are heavily correlated with the observed signal timing plan and heavily affected by the queue behavior and thus cannot be directly used for signal timing optimization.

12.2.3 EXIT-TO-DOWNSTREAM WAVEFORM PREDICTION MODEL

Exit-to-Downstream Waveform Prediction Model (M_{down}), this model predicts the modification of the exited waveform as it travels downstream to the next intersection. Given an exit waveform, the model aims to predict the waveform at a certain distance downstream from the intersection (not affected by downstream signal state). This waveform is treated as the inflow waveform to the downstream intersection.

Instead of predicting the waveform at downstream advance and stop bar detectors, we employ a two-step strategy: use the exit waveform to predict the downstream inflow waveform and use the predicted inflow waveform to predict waveforms at downstream advance and stop bar detectors. In this way, we can model the interactions between any pair of intersections independent of the distance between them.

12.2.4 STOP-BAR-ADVANCE WAVEFORM MODEL

Stop-Bar-Advance Waveform Model (M_{sa}), This model predicts the waveforms at stop bar and advance detectors, given the unperturbed inflow waveform and the signal timing plan. We can also use this model to verify that the inflow reconstruction done by M_{in} is of sufficient quality to be used to replicate the same observed stop bar and advance waveforms.

The waveforms are represented as 1-D vectors, each with T components. Here T refers to the length of time a particular sensor's data is being considered, with each component being aggregated at a 5-second level. In our work, T = 150, i.e., each data vector corresponds to 750 seconds of data (roughly 6-7 cycles), aggregated at the 5-second level. The time gap between two vehicles (headway) near an intersection is usually 2 seconds, which leads to an average of 2-3 vehicles per 5-second interval.

Table 12.3
Table describing driving behaviour parameters for Krauss car following model

Name	Description	Range
minGap	Min Gap when standing in (m)	[1.0, 5.0]
Accel	The acceleration ability of vehicles in m/s^2	[1.6, 4.6]
Decel	The deceleration ability of vehicles of this type in m/s^2	[3.0, 6.0]
sigma	Driver imperfection (how often sudden "bursts" of accel followed by decel	[0.1, 1.0]
tau	The driver's desired (minimum) time headway (sec)	[0.1, 3.0]

We find this level of aggregation sufficiently expressive to capture platoon dynamics and at the same time not overly compute-intensive to train our models. Signal timing information is encoded using eight vectors, each with T components, all either 1s or 0s, 1 indicating that a particular direction is green at that time interval (a typical intersection has eight phases or directions of vehicular movement).

As mentioned earlier, we use simulated data for training neural network models to predict these waveforms at a subcycle level. The main reason for using simulated data is that many unobserved quantities like queue lengths or waveforms at exit detectors or outflow detectors can be captured.

12.3 PROPOSED NEURAL NETWORK ARCHITECTURES

In this section, we describe the architecture of the Dual Attention Encoder-Decoder model. Figure 12.4 shows the proposed architecture. The model has four components: encoder, decoder, temporal attention module, and phase attention module. The encoder takes the waveforms at all the stop bar and advance detectors and generates a hidden representation. The decoder outputs the value of the output waveform at each time step. The attention modules help the network to concentrate on relevant temporal and spatial information. The architecture is described below.

The encoder is a GRU layer with 50 hidden units; input to it is of the size (*batch_size x no._of_input_variables* × *no._of_time_intervals*). In the forward pass, the last hidden

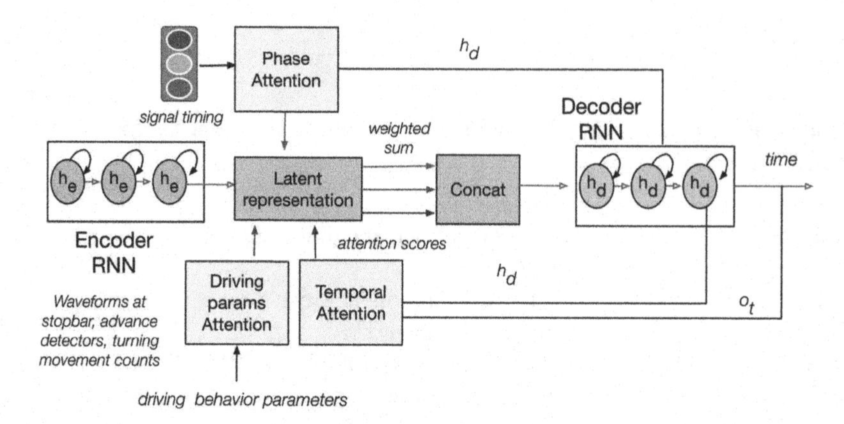

Figure 12.4: Dual Attention Encoder Decoder model (DA-ED). The Encoder RNN unrolls over the input waveforms to generate a hidden representation. The decoder is also an RNN that outputs the value of the output waveform at each time step, the input being encoder's output for the current timestep multiplied by the attention scores. Attention modules allow the model to focus on specific parts of the encoder outputs for prediction at current timestep. The temporal attention module is a feed-forward layer using the decoder's output at current timestep and hidden state as inputs. The phase attention module, driving attention modules generate attention scores which represents the importance given to each of the hidden units based on signal timing, driving behaviour parameters respectively.

state of the encoder($batch\_size \times 50$) along with all hidden states ($batch\_size \times 50 \times no\_time\_intervals$) is returned. The last hidden state of the encoder is used to initialize the initial hidden state of decoder. If only the last hidden state is passed through the decoder, it has the burden of representing the waveform across all the time points.

Attention modules allow the model to focus on specific parts of the encoder outputs based on the decoder's outputs. The temporal attention module is a feed-forward layer using the current decoder output and hidden state as inputs, and the output is a vector, ($batch\_size \times no.\_of\_time\_intervals$), which comprises attention scores representing the importance of each hidden state of the encoder for the current prediction. The phase attention module is another feed-forward network that uses the signal timing vector and decoder hidden state as inputs to generate attention scores. The phase attention score represents the importance given to each of the hidden units. These attention scores are multiplied by encoder outputs to create a weighted combination and then passed through the decoder.

The decoder is also a GRU layer, taking the weighted combination from both the attention modules as input ($batch\_size \times no.\_of\_time\_intervals{+}50$). The decoder outputs a prediction for the next time point and updates its hidden state. This predicted value is used by the attention module to generate attention scores for the next time step.

This architecture was inspired by recent advances in sequence-to-sequence models, namely transformer networks. Transformer networks also broadly use the encoder-decoder paradigm with attention. However, they trade RNN-based (GRU-based) encoders and decoders for convolutional neural networks (or feed-forward networks) to enable easy parallelization. Transformer networks are uniquely suited to natural language processing tasks, modeling long-range sparse dependencies such as correct subject-pronoun matching. For example, consider the sentences "The cat is sleeping on the mat. It has eaten its food." The word "it" in the second sentence depends squarely on "cat" in the first sentence, but not on "mat," which directly precedes it. However, in traffic waveform estimation, it is not the case that an event (such as a sudden inflow spike) occurring several cycles ago will suddenly affect the present cycle without having affected the intervening cycles. The congestion caused by the spike will dissipate over several successive cycles, and its effect will be contained in those successive cycles. Given that traffic state change is a gradual process, RNN architectures are well-suited for such situations, as they rely on the preceding hidden state for predicting the current state.

We compare the performance of the proposed architecture with feed-forward networks in section 12.4. The feed-forward network we use is a standard fully connected network with four hidden layers with 72, 56, 56, and 72 hidden units. We show that the proposed architecture has a much better prediction accuracy with 70% fewer parameters.

12.4 EXPERIMENTAL RESULTS

In this section, we present the experimental results for different models proposed in section 12.2 and compare the performance of the proposed architecture with standard feed-forward networks.

Even though the models are trained with mean square loss as the loss metric, we analyze the error in terms of *veh per bucket*. As our bucket is 5 s, the error denotes the absolute value of difference between actual and predicted number of vehicles in the 5-s bucket. Figure 12.5 shows the actual vs. predicted waveform for M_{exit} at different resolutions. The key observation here is that even though the actual vs. predicted waveforms may not exactly match at a 5-s-bucket resolution, if we aggregate them to higher resolution (10, 15, 25, 50 s, etc.), the actual and predicted waveforms almost match. Suppose the actual values in the next two buckets are 3 and 3, but the model predicts them to be 2 and 4; the error is almost 33% per bucket, but at 10-s resolution, the error is 0%. The error at different resolutions indicates that the overall momentum of the system is conserved (the total volume of the predicted waveform is equal to total actual volume). So we also report the error for different resolutions of the waveform even though our prediction is for 5-s resolution. Figure 12.6 shows the plot showing errors for predicting exit waveform using different models (M_{exit}). This plot shows that dual attention encoder decoder model has the best prediction accuracy.

Figure 12.6 shows the errors at different resolutions for M_{exit} using different architectures: feed forward, temporal attention, and dual attention. As mentioned earlier, the error is reported in terms of the number of vehicles per bucket. It can be

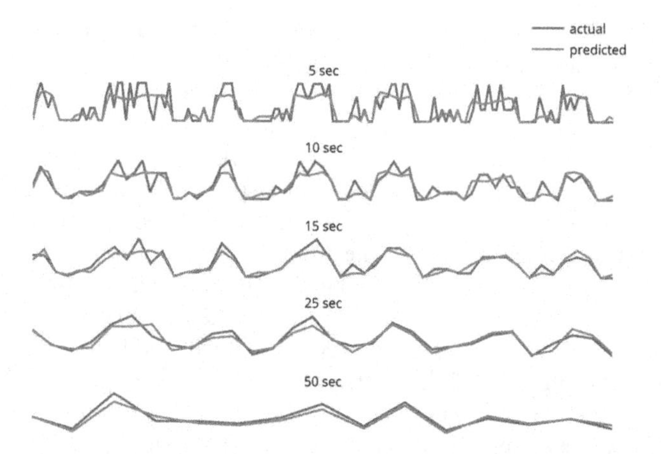

Figure 12.5: Plot showing actual vs. predicted exit waveform at different resolutions. The key observation here is that even though the actual vs. predicted waveforms may not exactly match at a 5-s-bucket resolution, if we aggregate them to higher resolution (10, 15, 25, 50 s, etc.), the actual and predicted waveforms almost match.

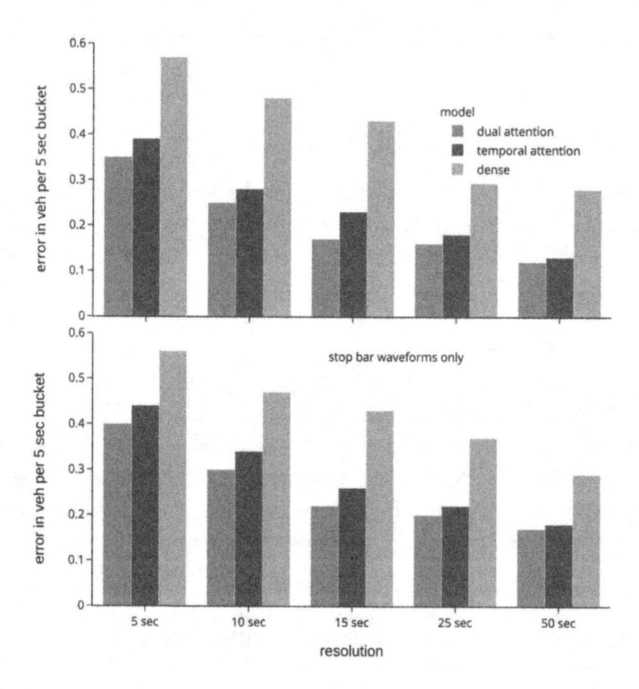

Figure 12.6: Plot showing errors for predicting exit waveform using different models (M_{exit}). This plot shows that dual attention encoder decoder model has the best prediction accuracy. A) Using stop bar and advance detector waveforms; B) Using stop bar detector waveforms only

Table 12.4

Table comparing errors for M_{down} model at different resolutions. The results suggest that using Robertson model followed by Dual Attention Encoder Decoder (DA-ED) model has better prediction accuracy. DNN: Deep Neural Network

Robertson	E, d	0.62	0.52	0.46	0.37	0.23
DNN	$S(E, d), d$	0.55	0.46	0.40	0.34	0.25
DNN	$R(E, d), d$	0.55	0.46	0.40	0.34	0.25
DA-ED	$S(E, d), d$	0.40	0.28	0.21	0.15	0.09
DA-ED	$R(E, d), d$	0.38	0.25	0.19	0.13	0.08

clearly seen that the dual attention encoder decoder (DA-ED) architecture outperforms the baseline feed forward network. Also, we can see that using only waveforms at stop bar detectors and signal timing also gives similar prediction accuracy.

Table 12.4 shows the errors at different resolutions for M_{down} model. This model predicts the modification of the exit waveform as it travels downstream. The table shows the error for predicting the downstream waveform using different models.

This model employs a two-step strategy to predict the downstream waveforms at a distance x:

1. First, the exit waveform is shifted in time, assuming average vehicle speeds to cover the distance x. To shifted exit waveform is used to impute the waveform at distance d based on platoon dispersion. We tried out two different strategies (1) Shift the exit waveform based on distance (d) - $S(E, d)$. This assumes that the waveform seen at the exit detector is largely unperturbed as it progresses downstream. (2) Use Robertson platoon dispersion model - $R(E, d)$.
2. Next, this shifted waveform is fed to a neural network model that modifies the waveform to incorporate non-linearities due to variable velocities, different driver behavior and signaling plan on the next intersections

It is important to note that the first step of shifting the waveform effectively captures the distance information between the two intersections. This allows the neural network model in the second step to focus on modifying the shifted waveform; thus, it is independent of the distance between the two intersections. Our experimental results suggest that using Robertson model followed by Dual Attention Encoder Decoder (DA-ED) model has better prediction accuracy

This model can be used to predict an inflow waveform for the downstream intersection. This inflow waveform is not affected by the signal timing state of either intersection. We can use the inflow waveform along with downstream signal timing information to reconstruct the waveforms at advance and stop bar detectors (M_{sa} model).

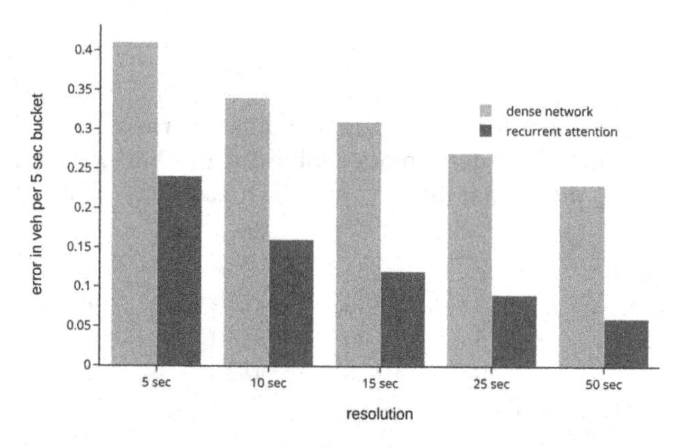

Figure 12.7: Plot showing error comparison for temporal attention encoder decoder vs. feed-forward model for inflow reconstruction. This plot shows that encoder decoder with attention model has the best prediction accuracy.

Figure 12.7 shows the errors at different resolutions for M_{in} comparing the dense network with temporal attention encoder decoder architecture. It can be seen that we could reconstruct the inflow waveform with good accuracy using stop bar and advance detector waveforms. At 50-s aggregation, the error in vehicles per bucket for the recurrent model is 0.06, equivalent to 0.6 vehicles (over- or under-counting by 0.6 vehicle for a 50-s period), whereas with the dense network, it is 0.27 or 2.7 vehicles. Figure 12.9 shows the plot of actual vs predicted waveforms for different driving behaviour parameters (X-axis: time, Y-axis: volume). The driving parameters (Table 12.3) are varied by keeping the inflow waveform and signal timing parameters fixed. This shows that the neural network models are able to capture the variation in driving parameters for modelling the waveforms.

Figure 12.10 shows the attention scores generated for each decoding step. The attention scores suggests that at each decoding step, more importance is given to hidden states of the encoder corresponding to that particular time step. For example, for predicting output at step 80, more importance is given to encoder hidden states corresponding to time steps 70-80. This suggests that the attention module is helping the model to focus on relevant temporal information at each step.

Also, It is worth noting that these methods are three to four orders of magnitudes faster than using microscopic simulations. For simulating input output patterns using SUMO, for a single intersection took 9 sec per simulation on a 32 core machine. While the trained neural network models when used in inference mode are able to generate output in less than a millisecond for a batch size of 5000. This suggests that neural network based model is atleast 4 orders of magnitudes faster compared to traditional simulation approaches. The models presented above M_{exit} and M_{down} can be stacked together to predict how the waveform changes as it travels downstream along the corridor. The M_{exit} is used to predict the waveform at the exit of the intersection and M_{down} is used to predict the waveform at the downstream intersection. The two models

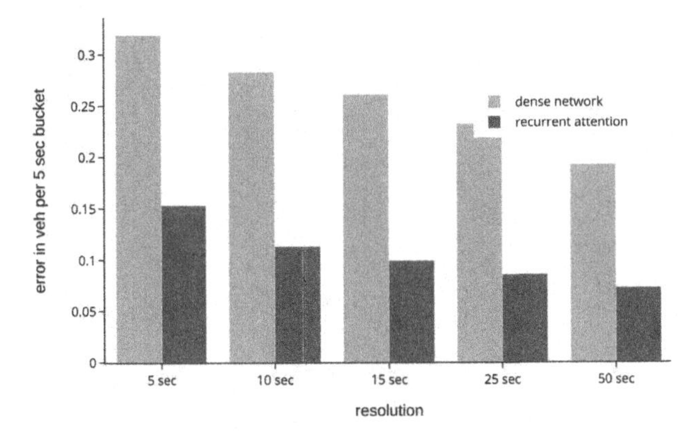

Figure 12.8: Plot showing error comparison for recurrent attention vs. feed-forward model for stop bar and advance waveform reconstruction. This plot shows that encoder decoder with attention model has the best prediction accuracy.

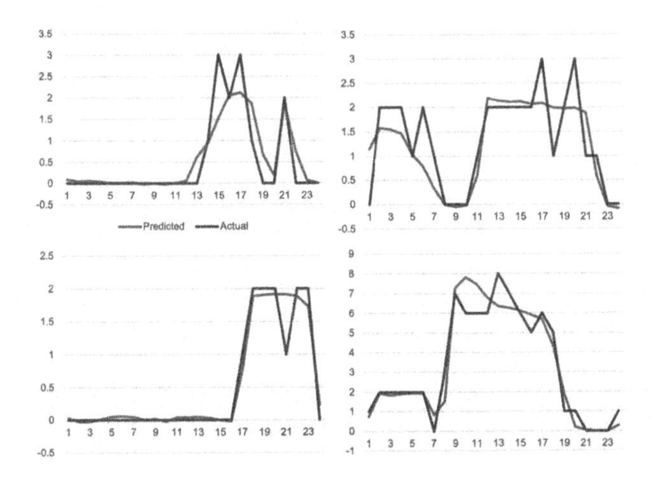

Figure 12.9: Actual vs predicted waveforms for different driving behaviour parameters (X-axis: time, Y-axis: volume). The driving parameters (Table 12.3) are varied by keeping the inflow waveform and signal timing parameters fixed. This shows that the neural network models are able to capture the variation in driving parameters for modelling the waveforms. Each of the subplot corresponds to different simulation exemplars with varied driving behaviour parameters.

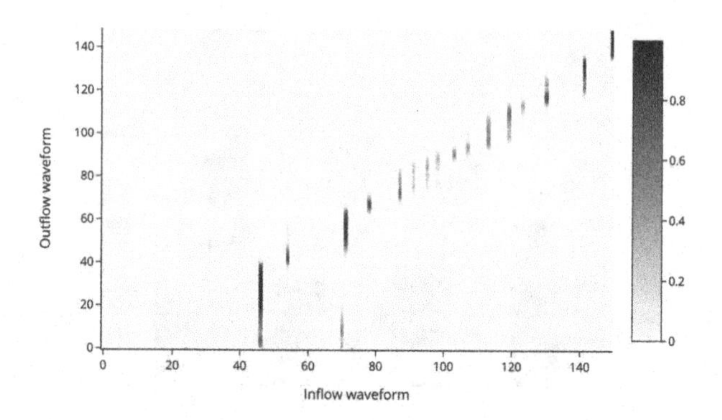

Figure 12.10: Plot showing heat map of temporal attention scores. The X-axis indicates the time buckets for the inflow waveform and Y-axis indicates time buckets for the predicted outflow waveform. This plot suggests that attention module is helping the network to focus on relevant temporal information, around the time for which the prediction is made.

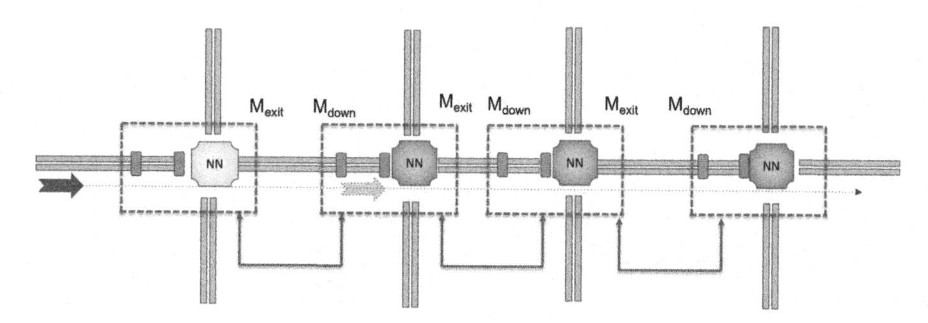

Figure 12.11: The models presented above M_{exit} and M_{down} can be stacked together to predict how the waveform changes as it travels downstream along the corridor. The M_{exit} is used to predict the waveform at the exit of the intersection and M_{down} is used to predict the waveform at the downstream intersection. The two models stacked across intersections would be useful to understand how platoons progresses across the corridor and compute level of service measures.

stacked across intersections would be useful to understand how platoons progresses across the corridor and compute level of service measures (Figure 12.11).

12.5 CONCLUSIONS

We described an approach for modeling waveforms at signalized intersections at subcycle resolutions. We arrive at the following conclusions:

1. We show that arterial traffic flow dynamics can be decomposed by considering local interactions (M_{exit}) and coupling interactions (M_{down}), and show

that they can effectively approximate the dynamics of a pair of intersections. An important advantage of using (M_{down}) is that the time-shifted input to the trained neural network model is independent of the distance between the two intersections.

2. We show how to effectively reconstruct the unperturbed incoming inflow waveform to an intersection from the stop bar and advance detectors and signal timing information waveforms (M_{in}). We verify that this reconstruction is accurate enough to reasonably estimate the observed stop bar and advance detector waveforms (M_{sa}). An important point to note is that the reconstructed inflow waveform is largely independent of the signal timing plan of the approaching intersection because it is still a significant distance away.
3. We present the accuracy measures of the model predictions and reconstructions improve with larger aggregation times, from 5 seconds to 60 seconds.

13 Modeling measures of effectiveness for intersection performance

Abstract - Microscopic simulation-based approaches are extensively used for determining good signal timing plans on traffic intersections. Measures of Effectiveness (MoEs) such as wait time, throughput, fuel consumption, emission, and delays can be derived for variable signal timing parameters, traffic flow patterns, etc. However, these techniques are computationally intensive, especially when the number of signal timing scenarios to be simulated are large. In this chapter, we present InterTwin, a Deep Neural Network architecture based on Spatial Graph Convolution and Encoder-Decoder Recurrent networks that can predict the MoEs efficiently and accurately for a wide variety of signal timing and traffic patterns. These methods can generate probability distributions of MoEs and are not limited to mean and standard deviation. Additionally, GPU implementations using InterTwin can derive MoEs, at least four to five orders of magnitude faster than microscopic simulations on a conventional 32 core CPU machine.

Several optimization methods have been successfully used to reduce travel time by optimizing the signal timing parameters. These methods try to derive the signal timing parameters such as green splits, cycle lengths offsets, etc. based on the traffic patterns for a given duration for a given intersection. A number of scenarios, each corresponding to a combination of different signal timing parameter values, is generated and simulation is performed to generate Measures of Effectiveness (MoEs) such as throughput and delays. MoEs are derived using macroscopic or microscopic simulation methods, the latter being more computationally intensive but generally more accurate. When these MoEs have to be computed for a large number of scenarios, the process can become computationally intensive. This is further accentuated when optimizing for a corridor or a network as the number of combinations increase exponentially. Methods such as ReTime address this by parallelizing VISSIM [89], SUMO [88], and AIMSUN [6]. Our approach does not require generation of trajectories and can directly compute MoE distributions. Once a distribution is available, statistics such as mean, variance, 90th percentile, etc. of an MoE can be easily calculated. We demonstrate the feasibility of achieving this for a single intersection with high accuracy. The main advantage of our approach is that when implemented on modern GPU-based processors, it is four to five orders of magnitude faster than running the microscopic simulators on a typical 32-core CPU machine. Clearly, if the actual vehicle trajectories are required on a given intersection, our method is not useful or appropriate.

In this chapter, we present an deep learning approach (called InterTwin as a short form for Intersection Twin) uses a large amount of data from different intersection topologies and signal timing plans, so as to capture the underlying traffic behavior at an intersection. We use microscopic simulators in conjunction with real world data to derive MoE distributions for a variety of scenarios that encompass a broad range of signal timing plans (as described in Chapter 6). In this chapter, we present the following

1. We describe a two-module deep learning approach that captures the intrinsic properties of traffic behavior at an intersection. The first module corresponds to a spatial graph convolution that is used to extract spatial features from the detector waveforms leveraging the relationship between intersection lanes and signal timing phases. This makes our modeling relatively independent of the intersection topology. The second module is an encoder-decoder with temporal attention architecture, to capture the temporal dynamic behavior of the traffic flow for each phase based on the signal timing plan. These two modules are stacked together for obtaining the final prediction.
2. We show that the InterTwin-trained models are able to accurately predict MoE distributions generated by traffic simulators. After training, when these models are used in inference mode, these models are four to five orders of magnitude faster compared to microscopic simulations. Additionally, it can model multiple intersection topologies without painstakingly redrawing a new base map for each intersection (that is typically required by a microscopic simulator).
3. For training these models, we use data generated using a significant extension of SUMO [88], an open source microscopic traffic simulator to make the data generation more realistic. We use real-world recorded data from high resolution loop detectors for input traffic patterns. Additionally, we have developed a new module that uses ring and barrier implementation along with arrival and departure information at the advanced and/or stop bar loop detectors along with signal timing information using techniques described in [67] and use them in our simulation to generate high fidelity MoEs that are reflective of data collected from real intersections. Additionally, we suitably vary signal timing parameters for these patterns to generate potentially viable counterfactuals. This results in our methods being able to generalize beyond what is typically used in actual practice and ensures that the models trained can predict robustly for a wide range of signal timing parameters. We also simulate a variety of intersection basemaps and behaviors, and estimate different measures of effectiveness, such as queue lengths, travel times, and wait times.

We present extensive experimental results to demonstrate the effectiveness of this approach. This includes comparison with other traffic-related deep learning models (that were not necessarily developed for predicting MoEs). We also show our model can be effectively used finding trade-offs between multiple MoEs at an intersection while evaluating different signal timing plans. Although the focus of this work is on a single intersection, we believe that the approach can be extended to corridors and networks and will be part of our future work.

The rest of the chapter is outlined as follows. Section 13.2 describes the proposed framework and architecture of the deep learning model presented. Experimental results are provided in Section 13.3, and a case study that uses this framework on a real world intersection is presented in Section 13.4, with conclusions presented in Section 13.5.

13.1 RELATED WORK

The idea of using deep neural networks to emulate physics-based simulations is not completely new. A deep learning framework with graph neural networks is proposed in [116] to simulate complex physical systems involving fluids, rigid solids, and other deformabale objects. This idea of building neural network-based emulators for physics-based simulations in different domains including high energy physics, climate science, astrophysics, and seismology has been explored in [70]. The authors propose an algorithm based on neural architecture search to approximate/emulate the simulations using deep neural networks that are accurate and orders of magnitude faster (up to 2 billion times). The main advantage of neural network emulators is that they are orders of magnitude faster and can be useful for extensive parameter exploration and very large scale analysis. Based on our detailed literature survey, we believe this is this is the first work proposing neural network emulators for traffic microscopic simulations to compute measures of effectiveness.

Machine learning techniques, including deep neural networks have been successfully applied to traffic state data in the past literature. Existing literature on predicting traffic state include either predicting either volumes or performance measures (such as travel times, wait times, queue length, etc.). The prediction horizon could be either at a sub-cycle level or at aggregate intervals (5 min to 1 h). We outline this work below.

A hybrid method incorporating filtering-based empirical mode decomposition is proposed in [141]. A deep learning-based method with non-parametric regression is proposed in [3], a novel deep neural network architecture with multisegents (recurrent and convolutional layers) is proposed in [140]. A deep generative model based on generative adversarial networks is proposed in [155]. A hybrid method based on linear programming, fuzzy logic, and multi layer perceptron is proposed in [161]. Some of the recent deep learning architectures proposed for spatio-temporal forecasting of traffic state data include temporal graph convolution network [164], spatio temporal residual graph attention network [37], and spatio temporal residual graph attention network [37]. Some relevancy to our work is proposed in [117, 10], which uses real world data to drive the microscopic simulations to compute performance indices such as travel time, emission performance, etc. that are used for the network. The main difference of this chapter from the existing work is two fold:

- We propose deep neural networks for estimating the distribution of performance measures instead of doing the microscopic simulations. Our methods are at least four to five orders of magnitude faster;

- Our models can also predict the performance measures for counterfactual signal timing plans, i.e., for a given input traffic and also for different cycle times and green splits (signal timing parameters). This can be useful to study the impact of different signal timing parameters.

The key observation is that neural network models can be used for computationally efficient parameter exploration. One important application in case of traffic intersections is that this approach can be used to find signal timing parameters for each of the intersection in a city corridor/network that satisfies a particular objective.

13.2 PROPOSED FRAMEWORK

We present deep neural networks to estimate performance measures such as wait times, travel times, etc., for a given input traffic and also for different cycle times and green splits (signal timing parameters). These trained models should be able to approximate the dynamics of physical simulators, when used in inference mode has the following advantages:

- The models are four to five orders of magnitude faster compared to simulations.
- The model can be used for multiple intersection topologies.
- These models can also be used to bootstrap the training of reinforcement learning-based optimization algorithms.

Effectively, the model captures the interrelationships between the various traffic, signal timing, and topology parameters and their impact on MoEs and allows for understanding the impact of a variety of signal timing parameters on MoEs for a given set of input traffic patterns for a given intersection topology. Additionally, our model generates a distribution rather than summary statistics. The overall workflow for training the neural network is shown in Figure 13.1. The data corresponding to different counterfactual signal timing plans is generated using microscopic simulators

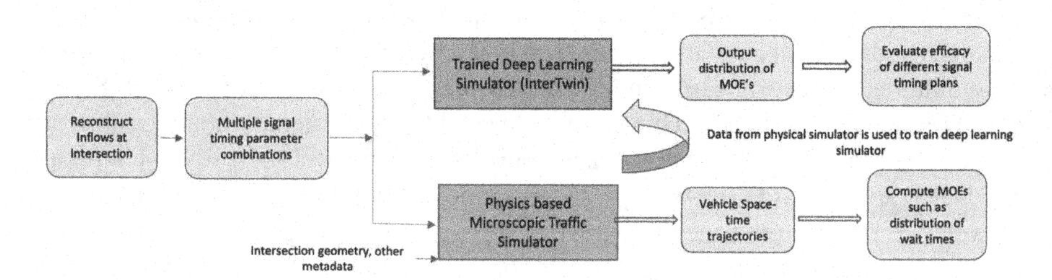

Figure 13.1: Overall workflow for training the neural network. The data corresponding to different counterfactual signal timing plans is generated using microscopic simulators to train deep neural networks. The trained neural networks can replace the simulators for predicting MoEs such as wait times and are four to five orders of magnitude faster. MoEs: Measures of Effectiveness.

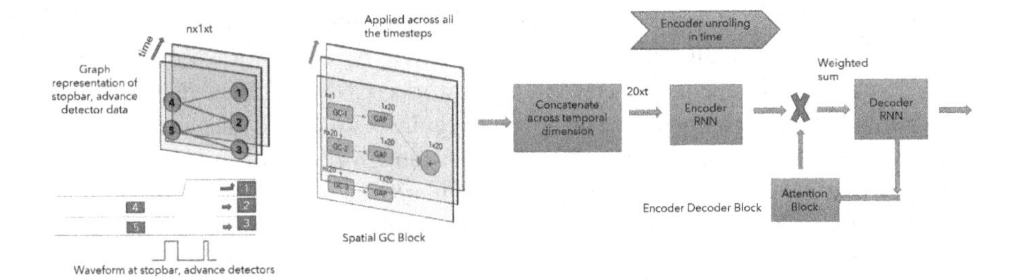

Figure 13.2: Architecture of the proposed InterTwin model. Spatial Graph Convolution (Spatial GC) is used to extract spatial features from the detector waveforms where the connectivity information of the intersection is incorporated. The encoder-decoder block is used to capture temporal dynamic behavior of the traffic flow.

to train deep neural networks. For the rest of this section, for ease of description, we use distribution of wait times as an example of our performance measure. However, our framework is general enough to be used for other measures (or multiple measures) as well. The deep learning model takes input traffic as input and outputs distribution of wait times for different signal timing plans.

The input traffic waveforms are represented as 1-D vectors, each with T components. Here T refers to the length of time a particular detector's data (arrival volumes) is being considered, with each component being aggregated at a 5-s level. In our work, $T = 72$, i.e., each data vector corresponds to 360 s of data. The output is the distribution of wait times for the 360-s window. The maximum wait time in our dataset is 2000 s, the wait times are binned into bins of size 10 s. So, the output is represented as a vector of size 200. Our proposed model, InterTwin, shown in Figure 13.2, has two main modules:

1. Spatial Graph Convolution (Spatial GC) is used to extract spatial features from the detector waveforms where the connectivity information of the intersection is incorporated
2. The Encoder Decoder with Temporal Attention (EDTAM) module is used to capture the temporal dynamic behavior of traffic flow. These two modules (GC and EDTAM) are stacked together for obtaining the final prediction. The details of each module is as follows.

We now describe each of these modules in detail.

13.2.1 SPATIAL GRAPH CONVOLUTION

The traffic movement on an intersection can be naturally represented in a graph structure, where each lane is represented as a node and an edge connects two nodes if one lane is feeding into another. We represent each detector (stop bar or advance) as a node and an edge exists between two nodes if there is a direct connection between them.

Graph convolutions provide a natural and meaningful way to extract features that can be used by higher layers where connectivity/spatial information has to be incorporated. Unlike standard Convolution Networks (CNNs) or Recurrent Networks (RNNs), graph convolutions can operate on irregularly structured data and can easily exploit the spatial structure of the intersection. Although CNNs are useful to generate adjacent spatial features, they are limited to regular/fixed grids).

In addition to being capable of handling different intersection topologies using graphs, representing an intersection as a graph structure can help the model learn from the spatial structure of the intersection, thereby providing relational inductive bias to the model. This also helps the model to generalize better for unseen intersections during the training phase [52]. Spatio-temporal graph convolution networks have been successfully used for city-scale traffic forecasting [160], where they show that their model outperforms other state-of-the-baseline models on a real world traffic dataset. Our approach models traffic at a much finer granularity as compare to the work in [160] (5 s versus 5 min).

Several methods have been proposed for generalizing convolutions on graphs [14], and they can be broadly classified into spectral-based or spatial-based methods. The spectral approach tend to capture the global structure of the graph more accurately than spatial methods. Unfortunately, spectral convolution methods require the eigen-decomposition of the graph laplacian, which is computationally expensive. There are several good approximation approaches available for spectral-based graph convolutions [73, 31, 79]. Considering the large size of our dataset, we employ the approach proposed by [73] based on first order approximation of localized spectral filters.

Consider an intersection is represented as graph $\mathbf{G} = \{\mathbf{V}, \mathbf{E}, \mathbf{A}\}$, consists of set of detectors $\mathbf{V}$ and $|\mathbf{V}| = n$, set of edges $\mathbf{E}$, and adjacency matrix $\mathbf{A}$. If there exists an edge between node i, node j then $\mathbf{A}(i, j) = 1, 0$ otherwise. Each node of the graph is represented as a vector $\mathbf{x} \in \mathbf{R}^t$, this corresponds to arrival waveform with t timesteps. Let $\mathbf{X} \in \mathbf{R}^{n \times t}$ be the node attribute matrix of the graph (n detectors, t timesteps). The notion of graph convolution operator in spectral graph convolution for a signal X with kernel Θ can be seen as:

$$\Theta_{*\mathbf{G}} X = \Theta\left(\mathbf{U}\Lambda\mathbf{U}^T\right) X \tag{13.1}$$

where $\mathbf{U} \in \mathbf{R}^{n \times n}$ is the matrix of Eigenvectors of the normalized graph laplacian; $\Lambda \in \mathbf{R}^{n \times n}$ is the diagonal matrix of eigenvalues. As discussed earlier, we use approximate methods to compute Θ. The convolution layer can be formulated as:

$$\mathbf{X}^{p+1} = \sigma\left(\tilde{\mathbf{D}}^{-\frac{1}{2}}\tilde{\mathbf{A}}\tilde{\mathbf{D}}^{-\frac{1}{2}}\mathbf{X}^p\Theta^p\right) \tag{13.2}$$

where, $\tilde{\mathbf{A}} = \mathbf{I} + \mathbf{A}$ is the adjacency matrix of graph $\mathbf{G}$, $\tilde{\mathbf{D}} = \sum_{j=1}^{n} \tilde{\mathbf{A}_{ij}}$ is the diagonal degree matrix of $\tilde{\mathbf{A}}$, Θ^p is trainable weight matrix of layer p, and σ (.) is the activation function The above equation represents the layer wise propagation rule that makes up a single layer of the graph convolution. At a high level, this graph convolution operation aggregates the neighboring node features at each layer.

To summarize, we use this graph convolution layer to extract spatial features from the detector arrival waveforms. The proposed Spatial GC module consists of 3 stacked graph convolution plus Global Additive Pooling (GAP) layers. The feature map after each GAP layer is aggregated to obtain the final output of Spatial GC block (as shown in Figure 13.2). The weights of Spatial GC block are shared across all the timesteps, the output for each time step is concatenated across temporal dimension which in turn is fed into the EDTAM module.

13.2.2 ENCODER DECODER WITH TEMPORAL ATTENTION

The encoder decoder model is a variant of Recurrent Neural Network (RNN) that has well suited modeling temporal sequences. These networks process input sequences within the context of their internal hidden state ("memory") in order to arrive at the output, the internal hidden state is an abstract representation of previously seen inputs. Thus, they are capable of modeling dynamic contextual behavior. We use Gated Recurrent Units (GRU) [22] as our choice of RNN in our implementation. The proposed EDTAM model consists of three building blocks—encoder, decoder, and temporal attention module. These are described below.

The encoder is an RNN that reads each timestep of the detector arrival waveform sequentially and updates its hidden state conditioned on the current input and its previous hidden state, Equation (13.3).

$$\mathbf{h}_{\mathbf{e}\langle\mathbf{t}\rangle} = f\left(\mathbf{h}_{\mathbf{e}\langle\mathbf{t-1}\rangle}, \mathbf{y}_{\mathbf{t}}\right) \tag{13.3}$$

where $\mathbf{h}_{\mathbf{e}\langle\mathbf{t}\rangle}$, $\mathbf{y}_{\mathbf{t}}$ are hidden state, input to the encoder at time step t respectively. The hidden state of the encoder is stored after each time step, and the final output of the encoder is $\mathbf{H}_{\mathbf{e}\langle\mathbf{T}\rangle} = \left(\mathbf{h}_{\mathbf{e}\langle\mathbf{1}\rangle}, \mathbf{h}_{\mathbf{e}\langle\mathbf{2}\rangle} \ldots, \mathbf{h}_{\mathbf{e}\langle\mathbf{t}\rangle}\right)$.

The decoder is also an RNN that is trained to generate the output sequentially based on its input and hidden state. At each time step, the input to the decoder is conditioned on the decoder's output at the previous timestep, encoder's output, and current hidden state of the decoder. The Temporal Attention Module (TAM) acts as an interface between the encoder's outputs and the input of the decoder.

The Temporal Attention Module (TAM) is a fully-connected network with the inputs being the decoder's output at a previous time step and its hidden state. The output of TAM is a vector, attention scores, which is used to compute the weighted sum of the encoder's hidden states. This weighted sum is fed as input to the decoder, the attention helps the model to focus on specific parts of the encoder's outputs for prediction at each step.

The final hidden state of the encoder is used to initialize the hidden state of the decoder. The attention scores for predicting at timestep, t are calculated as follows. Let $h_d\langle t\rangle$, z_t represent the decoder's hidden state, output respectively for time step t

$$A_T = \frac{\exp\left(\mathbf{w}_j\left[h_{d\langle t\rangle} z_{t-1}\right]\right)}{\sum_{j'=1}^{K} \exp\left(\mathbf{w}_{j'}\left[h_{d\langle t\rangle} z_{t-1}\right]\right)} \tag{13.4}$$

where w_j are the rows of learnable weight matrix $\mathbf{W}$, and A_T represents the computed attention scores. The input to the decoder is the dot product of these attention scores with the hidden states of the encoder, $A_T.H_{e\langle T\rangle}$.

13.2.3 OVERALL NETWORK

The Spatial GC block and encoder decoder module are stacked together and trained end to end with a chosen loss metric, Adam optimizer. Implementation, training, and evaluation of the model was done using the PyTorch [110] library.

13.3 EXPERIMENTAL RESULTS

We now present the experimental results of this approach and compare the performance of the proposed architecture with some of the recent deep learning architectures proposed for spatio-temporal forecasting. In particular, we compare the performance of our model against the following architectures.

- FCN: Fully Connected Network;
- RNN-FCN: Recurrent Neural Network followed by a fully connected layer;
- T-GCN: Temporal graph convolution network for traffic prediction as proposed in [164];
- STGCN: Spatio Temporal Graph Convolution Network for traffic prediction, as proposed in [160];
- ST-RGAN: Spatio Temporal Residual Graph Attention Network as proposed in [37].

As mentioned earlier, the output of our model is the distribution of wait times, $f_{WT}(wt)$, for a given input traffic pattern, signal timing parameters. Since we are trying to predict the distributions, a softmax operation is applied to the model output to convert it into a probability distribution. All the models are trained with the Adam optimizer and Mean Square Error (MSE) as the loss function. The different terms used to describe input, output variables is shown in Figure 13.3. Each of the waveforms aggregate counts for five-second intervals (the level of aggregation was chosen so that the number of vehicles in each interval is very small—generally less than three, but also large enough to keep the size of the network to be small).

Table 13.1 shows that our methods are better for all the error measures: Root Mean Square Error (RMSE), Mean Absolute Error (MAE), and Mean Square Error (MSE). These results show that InterTwin is significantly better than FCN, STGCN, and T-GCN in terms of mean square error for MoE prediction. The key advantage of predicting distributions is that different summary statistics such as mean, median, and percentile values can be derived. Figure 13.4 shows actual vs. predicted scatter plot of the 50th, 70th percentile of wait time computed from the distribution. Figure 13.5 shows actual versus predicted distribution of wait time for a given inflow waveform, for different green time splits. These plots show that the models are able to capture the interrelationship between input traffic and signal timing parameters. We trained

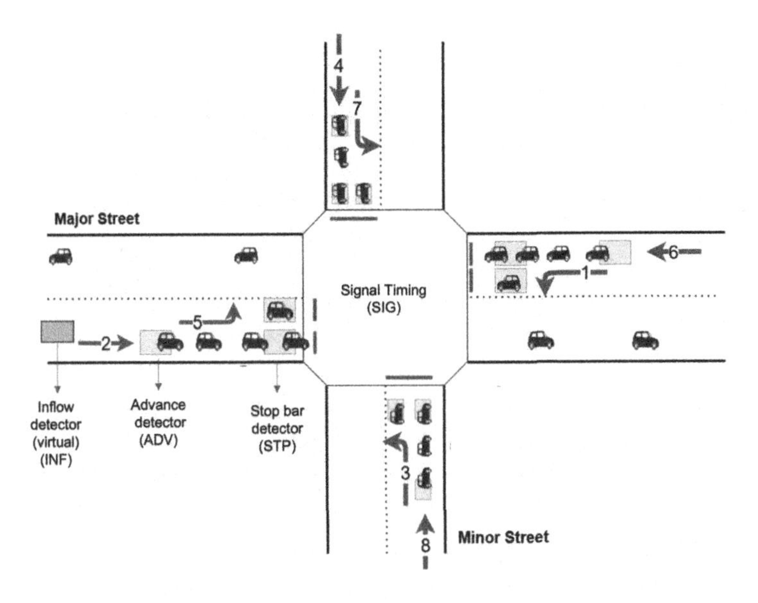

Figure 13.3: A typical intersection with 8 different directions of vehicular movement (phases 1 to 8). Vehicle waveforms are observed at Stop bar and Advance detectors (STP, ADV). STP, ADV, and INF corresponds to the traffic waveform at stopbar detector, advanced and inflow (500 m away from the intersection) aggregated at a 5-s interval. SIG corresponds to signal timing at a 5-s interval. STP, ADV, and SIG are typically available in ATSPM data for every intersection.

separate models each requiring different input traffic patterns for different Signal Timing waveform (SIG) using:

1. Only Inflow Traffic Waveforms (INF);
2. Only Advanced and Stopbar Waveforms (ADV and STP).

Table 13.2 shows the training and test errors for models with different input combinations. The accuracy is high when stopbar, advance waveforms are used as input compared to using only an inflow waveform. An important practical advantage of the model with only stopbar and advance waveforms as input is that both of these waveform are easily available in recorded controller logs at each intersection that support ATSPM. Thus, this model can be very effectively used to infer wait time distributions from recorded controller log data in real world. Unfortunately, this model cannot be used directly for computing MoEs for different signal timing plans. The latter is generally required for optimization purposes.

For optimization, it is more practical to use the INF waveform and SIG waveforms as input as this model allows for varying signal timing plans. As discussed earlier, INF waveforms for each approach can be computed using neural networks that use

Table 13.1

Table comparing performance of different models. The input to the models are ADV, STP, and SIG. This suggests that our InterTwin model has better accuracy compared to other model architectures. MSE: Mean Square Error, RMSE: Root Mean Square Error, and MAE: Mean Absolute Error.

Model	MSE	RMSE	MAE
FCN	1.5e-4	0.0084	0.003
RNN-FCN	2.2e-4	0.0091	0.0032
T-GCN [164]	1.2e-4	0.008	0.0026
STGCN [160]	1.5e-4	0.0085	0.0031
ST-RGAN [37]	5.4e-3	0.0165	0.0095
InterTwin (ours)	0.9e-4	0.0076	0.0023

ADV and STP waveforms along with SIG waveforms. For a given inflow, we can generate multiple candidate signal timing plans, use the trained model in inference mode to evaluate each timing plan based on the distribution of wait times. Using these trained models for MoEs estimation is highly scalable compared to simulation-based approaches.

For a given input traffic flow, to simulate 3200 different signal timing parameter combinations took more than 13,400 s on a 32 core machine. While the trained neural network model when used in inference mode generates output in 0.08 s for the same number of combinations (batch size 3200). This suggests that neural network models are at least four to five orders of magnitude faster.

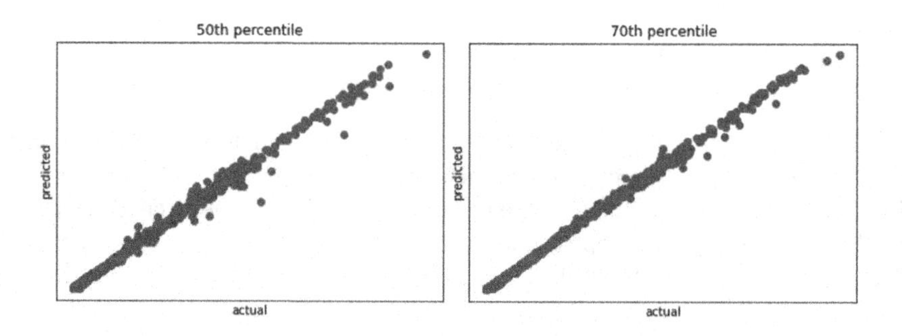

Figure 13.4: Summary statistics such as 50th/70th percentile etc. can be computed from the predicted distribution. Actual vs. predicted scatter plot of 50th, 70th percentile of wait time computed from the distribution. The key advantage is that predicting distribution enables us to compute any statistic of interest.

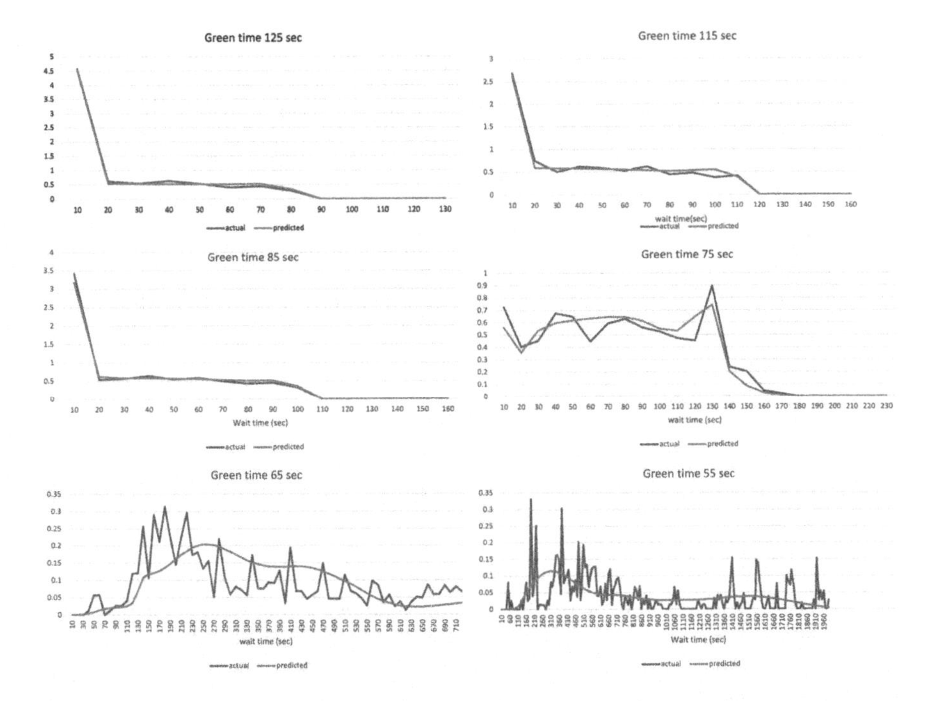

Figure 13.5: Actual vs. predicted distribution of wait time for different green time splits for a given input traffic. These plots show that the model is able to capture the interrelationship between input traffic and signal timing parameters.

Table 13.2

Comparison of model performance for different input parameters. This suggests that using STP, ADV has better accuracy compared to using INF waveform. The InterTwin model also has better accuracy. It is more practical to use INF along with SIG as INF waveforms are not affected by SIG and multiple signal timing parameters can be evaluated in parallel. Whereas, the other model (STP ADV SIG) can be useful to understand performance measures on recorded historical data.

Model	Inputs	Train Error(MSE)	Test Error(MSE)
InterTwin	STP ADV SIG	0.9e-4	0.9e-4
InterTwin	INF SIG	2.0e-4	2.1e-4
FCN	STP ADV SIG	1.3e-4	1.5e-4
FCN	INF SIG	3.0e-4	3e-4

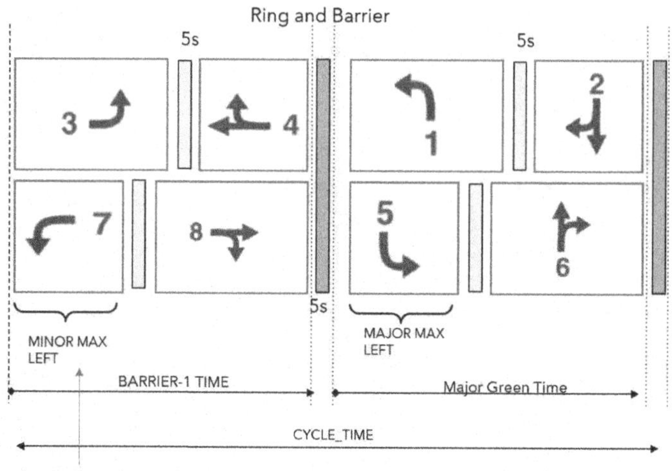

Figure 13.6: Ring and barrier controller allows us to separate the 8 lane-movements (i.e., phases 1–8) into two concurrency groups, with one group with the major street movements and one for minor street. Key signal timing parameters include cycle time, barrier time, max green splits, etc. For a given input traffic flow, the signal timing parameters are varied to generate different scenarios.

13.4 CASE STUDY

The trained models (that use INF and SIG waveforms) can be used to evaluate the efficacy of different signal timing plans for an intersection. This can then provide trade-offs for choosing a variety of signal timing parameters, including different green phase splits.

The 75th percentile of wait times as MoE is used for the rest of this discussion. In general, adding more time to the major street results in increased wait time on the minor street. The tradeoff can be clearly seen in Figure 13.7. It shows a scatter plot of 75th percentile of wait times for major vs. minor movements for different barrier-1 times (Figure 13.6). Based on this plot, a value of 60–70 s of green time may be appropriate as it minimizes the wait time on major street while not significantly impacting the wait time on the minor street. Of course a traffic engineer can look at these plots and other constraints to derive the optimal values. Given that this approach is computationally very inexpensive, optimal values can be derived separately for various hours of the day and day of the week combinations.

Table 13.3 shows a case study on a real intersection in Seminole County, Florida. We varied the barrier-1 time (keeping the cycle length fixed) to understand its impact on the wait time for traffic on major streets (NBT and SBT). These results show that changing the barrier time from 80 s to 60 s can lead to a 25% improvement in wait

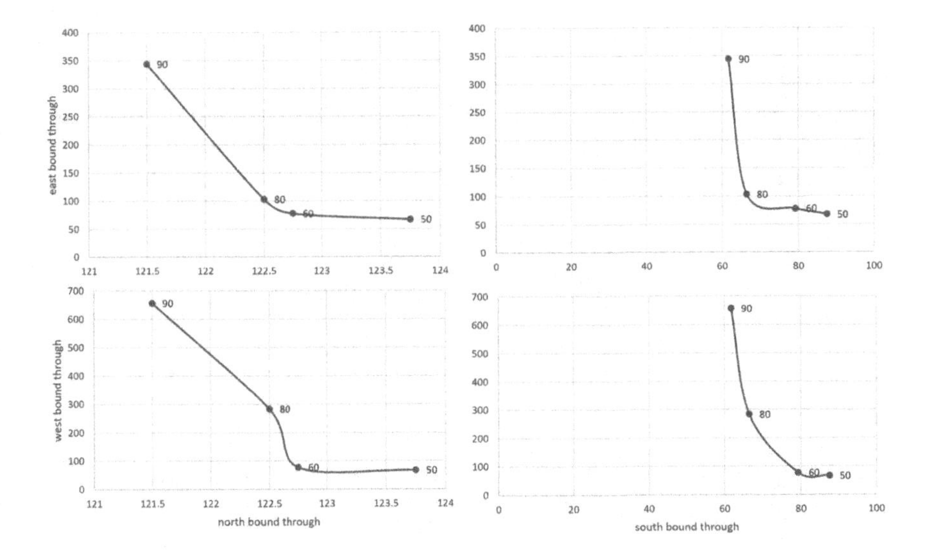

Figure 13.7: Scatter plot of 75th percentile of wait times for major vs. minor movements for different barrier-1 times. This can be useful to understand trade-offs in wait time for selecting different barrier-1 times for major vs. minor streets.

time for the major direction without significantly affecting the wait times on the minor street.

Rather than to provide one optimal signal timing plan, this framework can be extremely useful to practitioners to set green time splits for an intersection by understanding the trade offs for different hours of the day and days of the week.

13.5 CONCLUSIONS

In this chapter, we presented InterTwin, a deep neural network architecture based on spatial graph convolution and encoder decoder recurrent networks that can predict the MoEs quickly and precisely. Rather than just predicting one or two statistics for a MoE (e.g., mean and standard deviation), the network can compute the entire distribution. Additionally, these models are four orders of magnitude faster than conducting detailed simulations.

Broadly, we presented two models based on the input waveforms that are used along with signal timing. The first one uses stopbar and advance waveforms. An important practical advantage of the model is that both of these waveforms are easily available in recorded controller logs at each intersection that support an ATSPM-based system. Thus, this model can be very effectively used to infer wait time distributions from recorded real-world controller log data. This model, however, cannot be used directly for computing MoEs for different signal timing plans.

Table 13.3

Analysis on Intersection-1205 on 4 February 2019. Changing barrier-1 time from existing value 80 s to 60 s at 8:00 a.m. would improve the wait time on major direction by around 27%. NBT: North Bound Through, SBT: South Bound Through.

Time	Barrier Time New	Barrier Time Old	% Improvement-Wait Time-NBT	% Improve-mentof Wait Time - SBT
8:00 a.m.	60	80	26	29
9:00 a.m.	70	80	15	12
10:00 a.m.	70	80	12	10
11:00 a.m.	50	80	34	30
12:00 p.m.	60	80	21	21
01:00 p.m.	60	80	22	23
02:00 p.m.	60	80	22	24
03:00 p.m.	50	80	37	34
04:00 p.m.	60	80	25	33
05:00 p.m.	50	80	34	35

The second model uses only inflow waveforms as input as this model allows for varying signal timing plans. For a given inflow waveform, we can generate multiple candidate signal timing plans, use the trained model in inference mode to evaluate each timing plan based on the distribution of wait times.

We believe that this computationally-efficient approach can be extended to corridor optimization where the number of parameters is proportional to the product of parameters for each intersection on the corridor.

14 Signal Timing Optimization

Abstract: In this chapter, we discuss the field of computational signal timing . We discuss various mathematical modeling approaches, as well as machine learning approaches. We also discuss how signal offset optimization may be performed using a calibrated simulation model.

Isolated signals are relatively easy to optimize, and straight-forward algorithms that maximize flow given the state of vehicles queued at intersections are well-studied [163]. However, when the signals form a part of an arterial corridor, it becomes necessary to account for second-order effects i.e. impact of one intersection's signal timing plan on the upstream and downstream intersections. Also, in many practical scenarios, a grid can be decomposed into critical corridors, and these can be individually optimized. This may not always be possible, but it may be a reasonable compromise to simplify the problem. Thus, arterial signal timing optimization is an area of active research, with new technologies and sensor modalities being employed.

In general, these optimization techniques work by mathematically representing data pertaining to the state of the intersections, such as queue lengths, arrival patterns, etc. These are then used by the algorithm to compute an optimal (or near-optimal) signal timing plan for the corridor. Different constraints may be involved while arriving at the plan e.g. there may be minimum green times that the signal phases may have to adhere to. This signal timing plan is then implemented either in the real-world or in a simulator and performance measures are collected. Different performance measures that are used to quantify the efficiency of the signal timing plan include travel time, cumulative delay, number of stops etc. We now discuss different broad approaches to signal timing optimization.

14.1 MATHEMATICAL MODELING-BASED METHODS

Traffic flow theory [36] deals with the mathematical modeling of traffic flow phenomenon. While individual drivers may drive differently, based on their personal preferences and vehicles, overall certain broad trends can be represented mathematically. A thorough treatment of traffic flow theory can be found at [36]. The key variables that characterize traffic flow are:

- Flow: The rate at which vehicles pass a certain point (usually expressed in vehicles per hour).
- Density: The number of vehicles occupying a stretch of roadway (usually expressed in vehicles per kilometer).
- Speed: Time-mean speed (which is simply the arithmetic mean of vehicle speeds passing over a point) or space-mean speed (harmonic mean of vehicle speeds, represents the average speed over a length of roadway).

Among the first mathematical treatment of road traffic phenomenon, was the LWR (Lighthill, Whitham and Richards) theory [119] of traffic shockwaves. This theory has been widely used to study traffic phenomena on a variety of road geometries, including at signalized intersections, and signalized arterials. The field has grown greatly since then.

Several mathematical techniques have been proposed for optimizing traffic parameters such as offset, queue length, cycle length, phasing sequence and green split. For known traffic flow rates, it is possible to use methods like Webster's Method [145] to find optimal cycle time and green splits of all phases such that the average delay for all vehicles is minimized. Bandwidth optimization method uses traffic volumes, signal spacing, and the desired travel speed to determine the optimal progression band that can be achieved along a corridor. Examples include MAXBAND [85], MULTIBAND [123], PASSER (Progression Analysis and Signal System Evaluation Routine) [97], etc. Disutility-based methods minimize the "disutility" of a specific measure such as intersection delay or the number of stops. These models are generally designed to first find a common cycle length that minimizes the overall delay in the system, and then calculate the offset required for progression. Methods include TRANSYT (Traffic Network Study Tool) [150] and SYNCHRO [165].

The key idea of mathematical modeling-based optimization approach is to capture the dynamics of vehicles passing along the corridor, and using it to build an equation describing a Measure of Effectiveness (like delay, number of stops, etc. or a composite measure) [30, 86]. This equation consists of two components:

- Decision variables: Variables that a traffic authority can control, such as signal timing offsets, phase sequence, length of each phase etc.
- State parameters: Values acquired from the field (or intelligently-guessed using historical data) that describe the state of the corridor i.e. queue lengths, vehicle arrival platoons, vehicle speed limits, vehicle acceleration/deceleration etc.

This equation is subject to several constraints, such as minimum and maximum green times, cycle lengths, conflicting phase movements etc.

Then this equation is optimized using computational solvers [75, 55, 39]. Optimal or near-optimal results obtained can then be used to set the decision variables under the control of the traffic authorities.

We now briefly describe an optimization algorithm for single intersection and another for a corridor, to get an insight as to how such algorithms are designed. These serve to provide the reader an in-depth understanding of how such algorithms are designed, and are useful for understanding other such algorithms.

14.1.1 MINIMUM COST FLOW ALGORITHM

A Minimum Cost Flow(MCF) algorithm [112] for solving green time distribution at a multi-laned single intersection was developed for optimizing green times at a single intersection. The algorithm takes the number of detected vehicles at all incoming lanes as the input. Based on this, it sets up a minimum cost flow problem to move the maximum number of vehicles within the constraints of pre-decided phases (such as minimum and maximum green times per phase, yellow and red times).

The idea is to move as many vehicles as possible through the intersection given the constraints. Phases serving more lanes have a lower cost. Setting up the problem this way, allows us to leverage the well-explored field of Linear Optimization. Several computationally-fast Linear Programming Solvers such as CPLEX [9], GLPK [96] are available, which can solve the equations in a matter of seconds.

14.1.2 ARTERIAL TRAFFIC SIGNAL OPTIMIZATION

The integer programming-based algorithm outlined in [23] (called "Arterial Traffic Signal Optimization: A Person-based Approach") computes the optimal signal timing plan for a pair of intersections. It does so by setting up a large number of simultaneous linear equations that calculate the delay of vehicles at a pair of intersections, given a common cycle length and green splits for different phases. These linear equations build a mathematical model of the delay that the vehicles will experience, given a signal timing plan. Then, the algorithm iterates over all possible feasible plans to find an optimal plan that reduces the overall delay at the pair of intersections. The algorithm uses platoons of vehicles, while computing the delay, rather than continuous flows of vehicles. The algorithm requires the following inputs:

- Estimated Incoming Platoon Arrival Times
- Incoming Platoon Sizes
- Residual Queue Sizes

These can be estimated using various methods mentioned in the above sections. Then based on mathematical modeling of platoon flows, the delay time is calculated. This delay is the objective for minimization. Arterial travel time can be thought of as the combination of:

- Free-flow time: Travel time of vehicles crossing a fixed length of the corridor at maximum allowed speeds, assuming there are no traffic signals to interrupt the flow (i.e. like on a freeway).
- Delay time: Time wasted due to the presence of signaling at the intersection.

Since the free-flow time is constant for a particular stretch of the roadway, delay time is the key focus of optimization.

Once this information is provided, the algorithm searches through the various platoon break-up scenarios that may occur at a particular lane-group. These scenarios occur based on the size of the residual queue, and the size and arrival of the platoon in relation to the green phase.

The optimization is done pair-wise, in the direction of maximum traffic flow. In this way, the entire corridor is optimized. The policy obtained can be set until a change in the traffic pattern is expected (perhaps due to time of day or some other anticipated event). A new policy can be then computed and set.

14.2 MACHINE LEARNING-BASED METHODS

Recent advances in Intelligent Transportation Systems (ITS) have led to the widespread deployment of various real-time sensing and data collection systems such as high-resolution loop detectors, video cameras, GPS devices etc. Such streaming data has the potential for use in gaining insights into the state of corridors and real-time adaptive arterial traffic signal optimization [159, 163]. However, it is resource-intensive to collect, collate, store several terabytes of data, and then perform analytics to gain insights. Machine learning techniques, including Deep Neural Networks, have been successfully applied to traffic data for various applications[144].

Machine learning methods are usually supervised learning-based i.e. labeled data is fed to the learning algorithm, and the algorithm adjusts its parameters ("weights") to better predict the labels. However, in case of signal timing optimization, it is often not possible to obtain large labeled datasets. Hence, reinforcement learning is used, in conjunction with data generated in a traffic simulator.

Reinforcement Learning [126] is a field of machine learning in which a learning algorithm (an agent) interacts with a simulated environment and trains itself to accumulate rewards. Given a state of the environment at a certain time (the state can include past timesteps as well), an agent learns to take actions that maximize rewards (which are usually sparse and delayed). The actions an agent takes, based on the state input is called the "policy" of the agent. The agent interacts with the environment for several timesteps, in what is known as an "episode". An episode consists of a finite sequence of actions, state transitions and the associated rewards. An episode ends when a terminal state (like the agent "died" or "won" the game) is reached or if a predefined number of state transitions have taken place.

There has been a significant amount of research done in the field of applying reinforcement learning to traffic signal control. In Max-plus RL methods [99], agents (traffic intersections) calculate and exchange payoff values and use them to select their respective actions as part of the optimal joint action. In model-based approaches [72], agents create models of the operating environment or the action selection by neighboring agents and use that while deciding their actions. In actor-critic approaches [25], agents have their respective actor and critic functions, with the critic correcting the delayed rewards using temporal difference, which is then used by the actor to update the Q-values. In multistep backup reinforcement learning [62], agents update the Q-values based on the average effects of the temporal differences gathered in an episode, thus taking a view of the long-term effect of an action. [35] uses multi-agent reinforcement learning to jointly learn intersection behaviors.

In many cases, reinforcement learning is done with function approximation [40] i.e. agents represent Q-values using tunable weight vectors (like neural networks). Pairing reinforcement learning Techniques with recent advances in deep learning is the basis of deep reinforcement learning [82]. Deep reinforcement learning has been applied to the field of Traffic Control; [137] introduces a new reward function with a coordination algorithm for optimizing multiple neighboring traffic lights; [44] proposes a new discrete traffic state encoding for intersection states for use with deep reinforcement learning; [24] uses Multi-Agent Deep Reinforcement Learning (MARL) to coordinate traffic lights across a 5 by 5 grid of intersections.

14.3 QUANTUM ANNEALING-BASED METHODS

In recent times, there has been a great improvement in quantum processing technologies. Several large organizations such as IBM, Google, Rigetti, D-Wave etc. are building quantum computing devices. While these devices are still in their initial stages, they show great promise.

Broadly there are two lines of approach for these quantum devices:

- Universal Gate-Model Quantum Computing devices: In this class of devices, bits with quantum properties i.e. qubits, are used to perform quantum computing by applying operations (gates) on them. The promise of such an approach is that they are capable of doing everything a classical computing device can do and much more. However, these devices require highly accurate and stable qubits and gates, and the manufacturing technology currently isn't advanced enough to solve useful computational problems. Recently, Google announced that they had achieved "Quantum Supremacy", i.e. that they had demonstrated a computational solution on a quantum processor, that couldn't be easily solved by a classical supercomputer. However, the problem was extremely contrived, and there is still a long way to go before quantum processors decisively outperform classical supercomputers.
- Quantum Annealers: This class of devices uses a different approach to solving computational problems. In this model, qubits are used to create an energy landscape, which corresponds to the problem at hand. The system is then allowed to settle into a low-energy state, and sampling is done to get the results. While there are proofs that indicate that this model of quantum computing is equivalent to the Universal Gate-Model, the practical implementations of quantum annealers (such as those by D-Wave) do not aim to compete with them. For practical purposes, quantum annealers are used to find solutions to a specific class of optimization problems called QUBO (Quadratic Unconstrained Binary Optimization). While this may sound restrictive, a lot of important problems can be formulated in this paradigm. D-Wave quantum annealers have already been deployed by reputed research organizations such as NASA (National Aeronautics and Space Administration), Los Alamos National Laboratory [103] and Google Corp.

Quantum Annealing-based Optimization [98] is a computational technique for finding the global minimum of an objective function made of binary variables. Such a function can have potentially a large ($2^{Num\_Binary\_Variables}$) of possible solutions. But it is important to note that the solution space is finite. Thus, this formulation is ideal for decision-making problems, but for problems with infinite solution spaces, some degree of discretization may be required. For example, setting the speed of an aircraft under autopilot; speed is a continuous quantity.

D-Wave Systems has been developing devices for quantum annealing-based optimization for over two decades. These devices use quantum processes to perform the optimization. A D-Wave quantum annealer consists of a programmable array of "qubits" or quantum bits. D-Wave's 2000Q quantum annealer supports 2048 qubits, and D-Wave Advantage with 5640 qubits has recently been introduced [1].

A qubit can exist in the 0 or 1 state, or in a quantum "superposition" in between. Physically, a qubit's state is implemented as circulating current in a superconducting circuit. An external magnetic field acts as a "Bias", changing the probability of a qubit in superposition collapsing into the 0 or 1 state. Different qubits can be linked to each other, using the property of quantum "entanglement". This is achieved using "Couplers".

Using Biases and Couplings, the user can define an "energy landscape". This energy landscape is the problem the user wishes to solve. The D-Wave quantum annealer finds the minimum energy "valleys" of this energy landscape. These valleys correspond to optimal or near-optimal solutions. This technique is somewhat related to simulated annealing optimization technique, though Quantum Annealing benefits from quantum "Tunneling", which allows qubits to "tunnel" through classically-insurmountable peaks in the energy landscape.

QUBO [46] involves minimizing a Quadratic function of binary variables. This formulation is compatible with quantum annealers, with qubits playing the role of binary decision variables. The overall strategy of optimization using quantum annealers is to mathematically model the problem in terms of a QUBO, and add constraints in the form of penalty terms. For further details, please refer to [46].

In the paper [102] published by Volkswagen DataLab in association with D-Wave in 2017, one of the first attempts at solving an industrial problem in real-time using a quantum annealer is presented. This paper describes the formulation of a route-planning algorithm using QUBO paradigm to re-route vehicles in such a way to lower the route overlap and thus reduce overall congestion.

In[57], an approach to optimizing traffic grids is presented, which can also be used for corridors. The binary decision variables signify which phase at a specific intersection would be switched on for the next 5 seconds. The objective function is a mathematical formulation that maximizes the flow at each intersection but also tries to ensure that heavy flow phases are synchronized along the downstream route.

A method to globally control a large 50-by-50 intersection grid with 2-phase traffic is proposed in [58]. The formation of coalitions of agents in an interconnected system to coordinate actions, which can be applied to coordinating traffic signals in a grid is studied in [78].

Quantum annealing has been also applied to de-conflicting flight trajectories [124] and for the Capacitated Vehicle Routing Problem [38].

Offset optimization involves the use of computational methods to effectively coordinate traffic signal plans by varying the relative time offsets of the signal cycle plans

---

[1]https://www.dwavesys.com/press-releases/d-wave-announces-general-availability-first-quantum-computer-built-business

WE Volume: 869, EW Volume 1056 (With left-turn priors)

OFFSETS	WE_TT	EW_TT	Weighted Travel Time
[58, 30, 189, 9, 152, 149, 119, 114, 141]	389	264	320.4
[58, 23, 199, 33, 141, 151, 147, 121, 130]	382	271	321.1
[42, 26, 170, 18, 131, 146, 123, 108, 135]	383	271	321.6
[13, 26, 186, 24, 145, 143, 127, 123, 124]	386	272	323.5
Field Settings [36, 12, 176, 16, 156, 156, 129, 108, 126]	406	277	335.2

Figure 14.1: Optimal offsets found (in green rows) by the use of calibrated simulation models, over and above the deployed offsets (orange row). The top four rows are green, while the last row is orange. The colors are visible in our ebook version. The dominant flow is along the East-to-West (EW) direction with a volume of 1056 vehicles/hr, and the minor (yet significant flow) is West-to-East (WE) with 869 vehicles/hr. We optimize for volume-weighted travel time of both directions. We can see that new possible offset configurations with lower overall weighted travel times were found.

across the system (corridors and grids) in order to improve traffic flow and safety, while reducing stops, delays and incidents.

We have discussed traffic simulations, their calibration and deployment in the earlier chapters. We now present a real-world case study using a calibrated simulation model to find an optimal offset list for a nine-intersection urban arterial corridor during a weekday PM peak flow. We program the signal timing plan in the simulator based on deployed signal timing sheets for that duration (including the offsets and green splits). WEJO[2] sparse probe data along with left-turn traffic recorded by loop detectors was used to estimate the origin-destination matrix, in order to generate traffic. We use the parallel simulation framework (on the lines of the ideas discussed in earlier chapters) to find better solutions. The simulation model is simulated in parallel, and the travel time measures are recorded. While the offsets were perturbed at random, it is possible to use a guided search strategy such as genetic algorithm [71] etc. The top-performing offsets are presented in Figure 14.1.

---

[2]https://www.wejo.com/

15 Visualization of Traffic Data

Abstract - The advent of new traffic data collection tools such as high-resolution signalized intersection controller logs opens up a new space of possibilities for traffic management. In this chapter, we present visualization tools to provide traffic engineers with suitable interfaces, thereby enabling new insights into traffic signal performance management. The eventual goal of this chapter is to enable automated analysis and help create operational performance measures for signalized intersections while aiding traffic administrators in their quest to design 21st century signal policies.

The data available at each intersection can be broadly divided into signal timing data and vehicle detector data (gathered from loop detectors). The former consists of traffic movement and timing information for different phases, while the latter consists of arrival/departure and occupancy information for vehicles. In the past, this information was available at coarse levels of granularity (for example, traffic movement counts by the hour) limiting their use and the ability to discover cycle-by-cycle changes. Methods such as the Purdue Coordination Diagrams [28, 134] have been shown to be useful in understanding signal behavior and potential bottlenecks using signalized intersection datasets.

The availability of high resolution (10 Hz) controller logs opens a broader range of possibilities that were not available in previous systems. Additionally, this data, in many cases, is available with small latency (a few minutes) making it amenable to real-time decision making and addressing of bottlenecks. However, this plethora of information without proper decision-making tools adds a burden to transportation professionals. Many of them evaluate this information using ATSPM[1] tools and analyze the collected information one intersection at time. This is challenging even for small cities comprising of a few hundred traffic intersections. There is a need for a system that provides corridor-level and city-level information in a succinct and actionable form. In this chapter, we describe a system that partially addresses this need. This system leverages machine learning methodologies for data collected from a large number of intersections to derive key spatio-temporal traffic patterns in a city and then interactively allows a traffic engineer to focus on key challenges or improvements that can be carried out to alleviate them. Additionally, this system provides an analysis of traffic interruptions by observing changes in traffic at detectors, approaches, and intersections. The key modules of the system are now described.

1. Ranking: Several measures of effectiveness (MoEs) are used in the discipline [28, 135]. This system allow the user to rank or select intersections based on split failures and ratios of arrivals on red vs. green. Additionally, using a combination of these two metrics, we subdivide the intersections into several categories.

[1]`https://udottraffic.utah.gov/atspm/`

2. Clustering: Intersections with similar behavior or performance are grouped together using machine learning techniques. This approach is particularly useful when dealing with a large number of intersections and is carried out along both space and time. The system discovers and highlights signals on a corridor that preform similarly.
3. Change detection: We present a change detection algorithm that can detect statistically significant changes at an intersection level as compared to previous, similar time periods. This approach can be used to determine unexpected behavior or change in traffic patterns.
4. Incident detection: Using time series analysis, we derive extended time periods of significant traffic reduction for a detector or for an approach. A spatio-temporal presentation of this information is useful to derive key areas of traffic interruptions.

This system, which can be executed in parallel on a multicore machine and can handle data from thousands of intersections. The system can process six months of data for 300+ intersections (roughly 1 Terabyte) in less than 6 hours using a 50-core processor. A visualization module allows the user to select spatial and temporal regions of interest in an interactive fashion.

The rest of the chapter is organized as follows. The pre processing steps to convert raw controller logs to measures of effectiveness (MoEs) is already described in 8. In Section 15.1, we introduce the key modules in the system. These include: intersection ranking and categorization, clustering methods to highlight spatio-temporal patterns in the performance of intersections, change detection and incident detection. In the fourth section, we detail our visualization framework. In the final two sections, we detail the overall workflow, present results to demonstrate the performance & scalability of the system and summarize our contributions. We also discuss how our techniques can be incorporated into existing ATSPM systems.

The availability of high resolution (10 Hz) controller logs opens has open up a broad range of possibilities in terms of traffic intersection monitoring and performance metrics. The new generation of signal controllers, based on the latest Advanced Transportation Controller (ATC) [2] standards, are capable of recording signal events as well as vehicle arrival and departure events at a high data rate (10Hz). This allows us to compute signal performance metrics such as arrivals on red, arrivals on green, and platooning ratios on a cycle-by-cycle basis [135].

We can ingest data on a daily basis for the advanced controllers of type NTCIP 76.x, ATC. The detailed data collection process is described as follows. Each controller stores 24 hours of this data. This data is collected once a day using FTP by providing the IP addresses, which are local addresses to the remote network. A script can be used to initiate a FTP connection to each controller, downloads the stored data, decode the data to ASCII format, and upload the data to a local computer for further processing. The raw data is then processed, and the required information extracted and stored to a database.

[2]https://www.ite.org/technical-resources/standards/atc-controller/

15.1 FUNCTIONAL ARCHITECTURE & KEY MODULES

The overall approach seeks to process data and aggregate it at cycle-by-cycle level. It is worth noting that cycle times, in general, are variable throughout the day for each intersection. This cycle level data is then used to generate several measures of effectiveness that are further aggregated to fixed size intervals (e.g., 15 minutes or an hour).

15.1.1 RANKING AND CLASSIFICATION

We use demand-based split failures (computing red occupancy and green occupancy ratios) and the ratio of arrivals on red to arrivals on green (AoR/AoG) as measures of effectiveness (MoEs) of an intersection. These measures serve as good proxies for the level of traffic demand and effectiveness of signal timing, respectively. The possibility of using other measures is being explored further.

Once these measures are computed, they can be used for both filtering (using a threshold) or ranking. This allows traffic engineers to focus their effort on the most problematic intersections. A combination of the above MoEs is then used to categorize intersections into four broad classes:

1. Low split failures, Low AoR/AoG: Well timed and utilized intersections
2. Low split failures, High AoR/AoG: Low demand but potential for timing improvements
3. High split failures, Low AoR/AoG: Potential capacity problems
4. High split failures, High AoR/AoG: High demand and potential for timing optimizations

Additionally, intersections with detection issues or missing data can be derived if these measures are very high or very low for extended periods of time.

15.1.2 CLUSTERING

Daily data for each intersection based on MoEs is represented as a vector. The length of the vector is based on the number of intervals into which the entire day is divided. For example, if the data is aggregated at an hourly basis, the length will be 24. A weighted graph is first constructed for all intersection and day pairs based on distances between the vectors. Nonlinear dimensionality reduction, followed by clustering in the reduced space is then used to produce the clusters of similar behavior based on the MoEs. For more details the reader is referred to [92].

A representative set of cluster centers is derived and described in Figure 15.1. Sometimes, the intersections belonging to a cluster are spread over geographic regions several miles apart. While these intersections may be performing similarly, there is limited real value in having such distant intersections in the same cluster. A second round of processing is performed to split a cluster of intersections into multiple disjoint clusters based on spatial or corridor locality. A geographical indicator such as primary road names or distance between the intersections is used for this purpose.

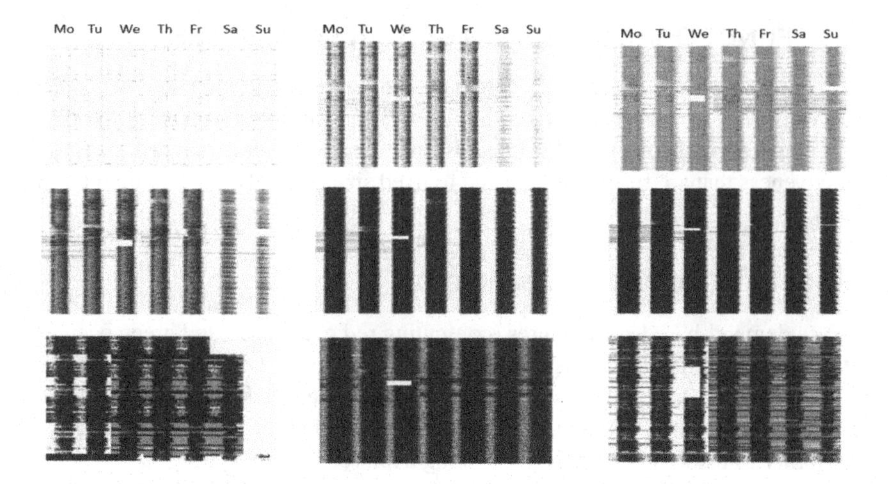

Figure 15.1: The nine clusters corresponding to distinct demand patterns discovered in the data. Each plot represents one cluster, and reach row represents the weekly behavior of an intersection that belongs to the cluster. We order these clusters from low demand to very high demand.

15.1.3 CHANGE DETECTION

This module can be used to discover significant changes in signal performance. Our methods detect temporal changes in signal performance, and/or detect periods with changes in many signals, and automatically highlight the change or evolution of intersection performance with time. The following approach is used:

- A significant change in the performance of the intersection over time can be detected by observing the evolution of the intersection's cluster membership over time.
- A change in the lower dimensional projection of the data representing the intersection performance can be used as a change detection measure.

In practice, the two methods are combined into one overarching method. Details are provided in [95].

15.1.4 INTERRUPTION DETECTION

Managing traffic interruptions is one of the crucial activities for any traffic management center. These interruptions may be caused by traffic accidents, vehicle breakdowns, debris, etc. Further, this should be done in (near) real-time so that proactive actions can be undertaken for mitigation. Broadly, we define an interruption to be any time period where the amount of traffic is significantly lower than normal or predicted traffic for a significant period of time. A *large* traffic interruption is defined on the bases of two parameters (see Figure 15.2):

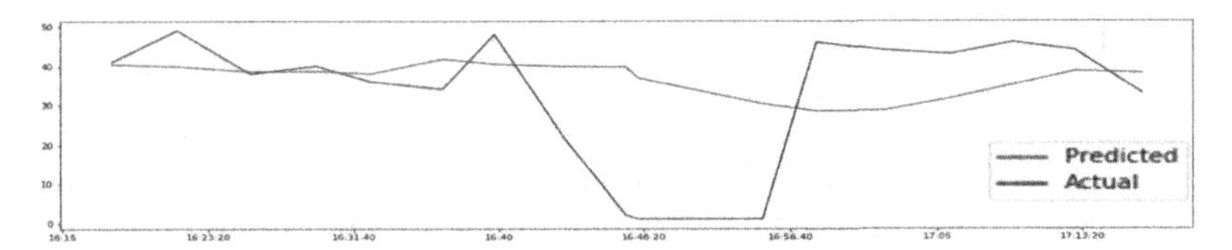

Figure 15.2: Example of a major interruption. Note the significant deviation of traffic volumes from predicted volumes (amount and duration).

1. The magnitude of deviation (percentage reduction) of observed traffic volumes from predicted volumes. This is measured in terms of the percentage dip of the actual traffic volume vs. the predicted value. Common sense dictates that the greater the deviation, the larger the interruption.
2. The duration (in seconds) for which the actual traffic volume is less than a *baseline* predicted volume. Again, a long duration heralds a large interruption.

The reader is referred to [64] for further details.

15.2 VISUALIZATION

We have developed a visualization module that allows the user to derive trends and hotspots in their city using the modules described in the previous section. The key features of this module are the following:

1. The user can select a small subset of intersections based on MoEs of interest. This is important as it allows the user to focus on problem areas. Each signal is represented by an icon based on the four categories derived by the combination of the two MoEs. The size of the icons, used to represent each of the different categories (based on the two MoEs), is proportional to the relative severity. This allows the user to visually compare the different intersections.
2. The user can hover on a particular intersection to get a detailed description of the MoEs and other relevant information at a granular level. Examples of such information include the signal ID, the number of split failures that happened on an hourly basis for the major approaches, the number of arrivals on red and green and the number of pedestrian actuations.
3. The intersections in a cluster with similar behavior can be easily identified because they have the same color. This allows the user to observe spatially and temporally similar behavior across multiple hours or days. Further, the user can highlight all the intersections represented by a particular cluster. For each cluster, the behavior legend presents the corresponding color, the number of members, the name of the road where the members may be found, and the days of the week that the cluster was observed.

4. There are different screens for each functionality. Further, different screens allow the user to access information for a single time period or multiple time periods. The former is useful in looking at details for a single time period while the latter allows the user to see comparisons between time periods or trends. The time periods for multiple period screens can be chosen by a drop-down menu, and the user can flexibly chose two to ten time periods.

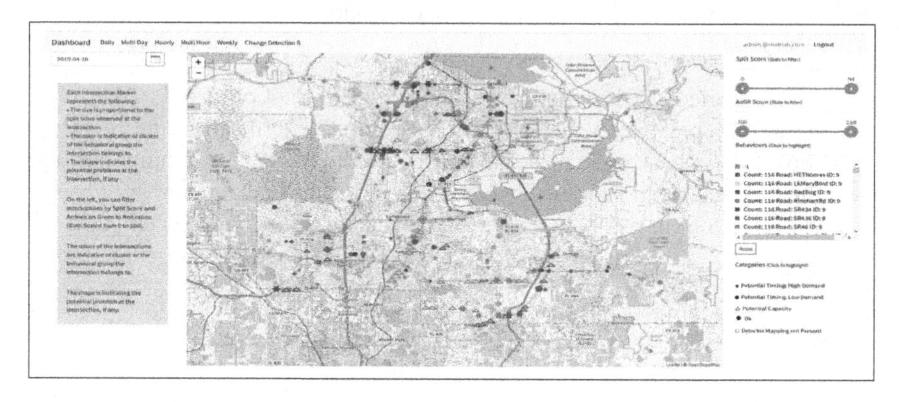

Figure 15.3: Dashboard showing the clustering and classification results for a single day. The results show that intersections on the same corridor demonstrate similar behavior throughout the day.

Figure 15.3 shows the results on a single day. Many signals on the same corridor get grouped together, showing that they performed similarly during the day. The clustering results of a multiday dashboard in Figure 15.4 show that for many intersections, the performance is similar during weekdays but differs from the weekends. For this particular week, many of the intersections performed well on Sunday but had potential capacity issues on weekdays.

The clustering technique is sensitive enough to separate intersections with granular differences between the observed behavior that are limited to only a few days. i.e. certain behavior can occur only on weekends whereas other patterns exist throughout the weekdays. Similar screens are available for performing this analysis on an hourly basis . Thus, the visualization system can be used to understand key behaviors in a grid or network of signalized intersections. It can be used to understand the hours or days for which the traffic patterns are similar and the time periods for which there might be some problems

Figure 15.5 shows the change detection screen of the dashboard. The user can select two time periods for comparison and the system provided differences in behavior. If only one time period is chosen, the system automatically chooses a baseline (e.g., for a Monday, it will chose the previous 6 to 10 Mondays). Statistically significant changes are then presented, allowing the user to detect temporal changes in traffic behavior.

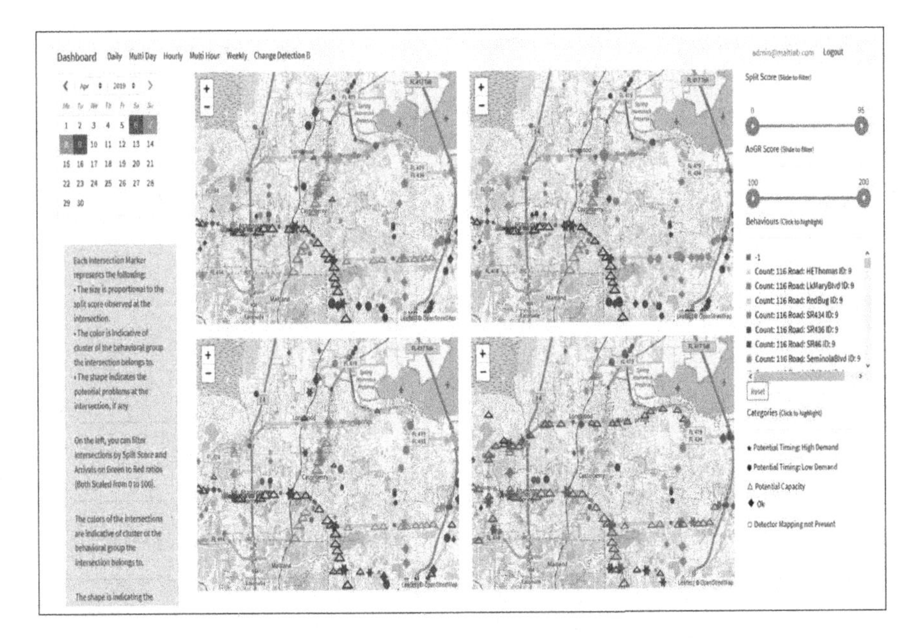

Figure 15.4: Multi-day angular dashboard allows for comparison of the clustering and ranking results across days. It highlights temporal patterns in intersection performance. We can see that the behavior on weekdays can be contrasted with the weekend behavior.

15.3 SYSTEM AND PERFORMANCE

Our system is implemented using a variety of software technologies based on Python, Elixir[3], and Angular[4]. We have also used libraries such as NumPy, Scikit-Learn and Pandas. A high level architecture of the system in presented in Figure 15.6. At the lowest level the monitoring module collects data from a large number of intersections. The module allows for performing collection in real time or at regular intervals. This data is stored in a database. A multi-threaded software layer based on Elixir and Python is used to develop all of our algorithm implementation for each of the modules described in the previous section. This allows for fault tolerance and seamless scalability in presence of additional computational resources. In particular, the architecture has several useful attributes:

- Fault Tolerant: We create a single actor for each functional task, and each actor or thread has it's own supervisor thread. If an actor fails to execute it's task, it suspends itself and all of its children and sends an exception to its supervisor. The supervisor can then work on a recovery strategy. Actors and

[3]Elixir: `https://elixir-lang.org/`

[4]AngularJS: `https://angularjs.org/`

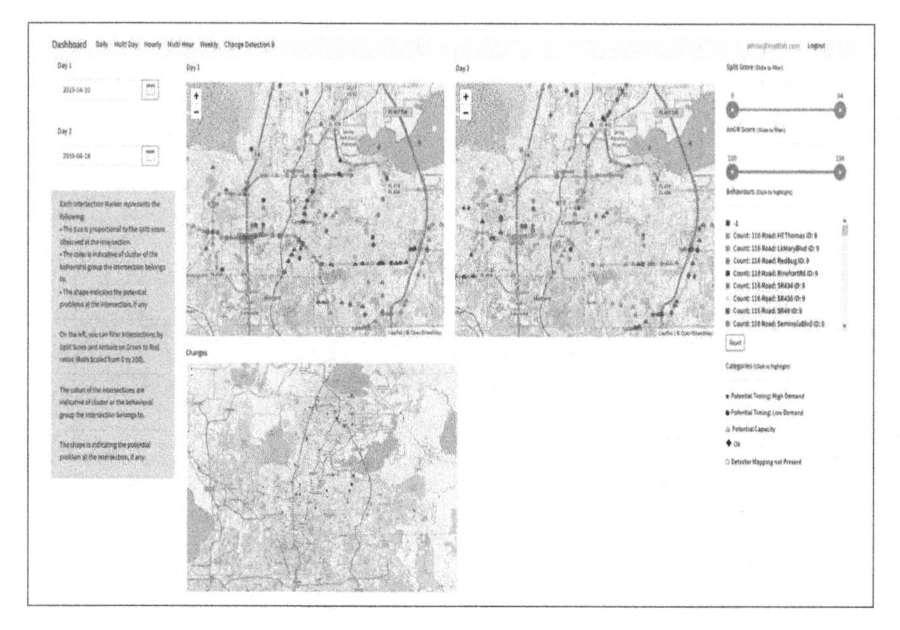

Figure 15.5: The Change Detection dashboard allows the user to compare and detect statistically significant changes between any two dates or from the baseline behavior.

supervisors fail gracefully, and all the failures can be ultimately managed by the Elixir/Erlang virtual machine (called BEAM).

- Scalable: Because the system has a set of dedicated actors for each functional module and the actors have very little overhead associated with them, it is easy to spawn new actors with an increase in the workload for any specific module. The number of actors is only limited by the resources available on the physical machine.
- Parallel and Distributed: This architecture is inherently parallel and can be distributed among multiple machines for further scalability. Any number of nodes running the virtual machine can be merged easily, and the actors or threads running on one node can communicate with threads on other nodes with no extra effort. The only additional overhead is the latency associated with communicating over a network.

The performance of the system was evaluated on a 50-core CPU server. 1 month of controller log data from approximately 1000 intersections was processed in a total of around 180 minutes. The workloads are a function of city size and we have documented batch performance observed on historic data. But, given the performance observed for the current (historic data based) workloads, this system can easily scale for near-real time workloads.

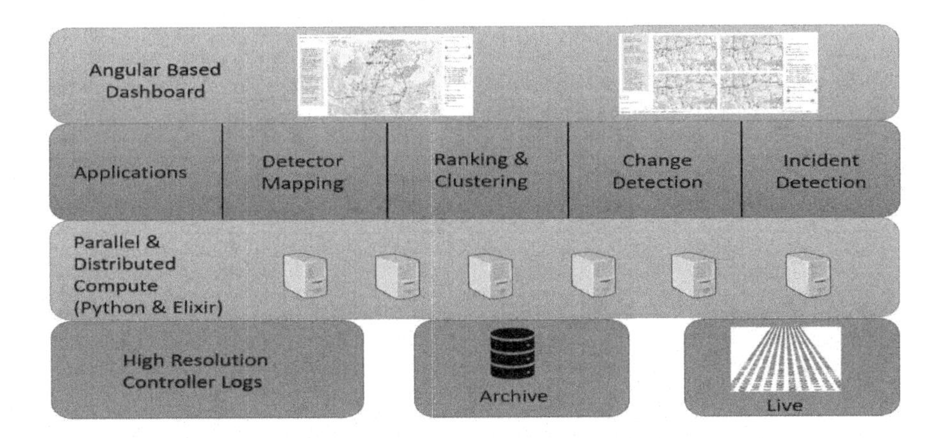

Figure 15.6: Overall Architecture of the system. starting with event logs from controllers, we use distributed and parallel compute techniques to design a robust framework to support various applications.

Figures 15.3, 15.4 and 15.5 show a snapshot of a visualization built in Angular 6 (with a Ruby backend) to present our results. The main benefit of choosing Angular is its component-based architecture, which enables the reuse of components and elements across the application. Also, the use of services in Angular assists in sharing the data across components with similar functionality.

The maps that are embedded in the application are built using leaflet.js. Leaflet is a JavaScript library that provides interactive maps and contains all the required mapping features. Since the application deals with a large number of intersections, it is important for both scalability and responsiveness to show these intersections as markers on the map without any noticeable lag. For example, whenever the user selects a range of dates, more than 10,000 markers are plotted on the maps. Making the markers are made with SVG or HTML reduces the performance of the application because all the markers have to be loaded into the DOM (Document Object Model). To overcome this issue, each marker is drawn using Canvas, and because Canvas markers need not be loaded into the DOM the maps can handle more than 100,000 markers at once without sacrificing user interactivity.

A Acknowledgements for Materials

We would like to thank Rohit Reddy, Varun Reddy Regalla for collaboration for portions of the work presented in this book.

We would like to thank Raj Ponnaluri, Trey Tillander, Darryll Dockstader, John Krause, Tushar Patel, Siva Srinnivasan and Lily Elefteriadou for providing encouragement and support throughout this work. This work was supported in part by the Florida Department of Transportation (FDOT), FDOT District 5 and NSF CNS 1922782. The opinions, findings, and conclusions expressed in this publication are those of the author(s) and not necessarily those of FDOT, FDOT District 5, or of the U.S. Department of Transportation and the National Science Foundation

We would like to also acknowledge the following contributions towards the chapters in this monograph.

1. Chapter 1 used materials copyrighted by IEEE. Specifically, the definitive version appeared in the proceedings of the IEEE 23rd International Conference on Intelligent Transportation Systems (ITSC), 2020 [93].
2. Chapter 6 used materials published by SCITEPRESS. The definitive version appeared in the proceedings of the International Conference on Vehicle Technology and Intelligent Transport Systems, 2021 [66].
3. Chapter 7 used materials from two sources. The definitive version of the clustering work appeared in the proceedings of International Conference on Vehicle Technology and Intelligent Transport Systems, 2020 [91].
4. Chapter 8 used materials from three sources. The definitive version of the summarization work appeared in the IEEE 23rd International Conference on Intelligent Transportation Systems (ITSC), 2020 [95], and was copyrighted by IEEE, while the definitive version of the data driven approach appeared in the proceedings of International Conference on Vehicle Technology and Intelligent Transport Systems, 2020 [91].
5. Chapter 9 used materials copyrighted by IEEE. Specifically, the definitive version appeared in the proceedings of the IEEE Fifth International Conference on Fog and Mobile Edge Computing (FMEC), 2020 [65].
6. Chapter 10 used materials published by SCITEPRESS. The definitive version appeared in the proceedings of the International Conference on Vehicle Technology and Intelligent Transport Systems, 2021 [66].
7. Chapter 11 used materials copyrighted by IEEE. Specifically, the definitive version appeared in the proceedings of the IEEE International Intelligent Transportation Systems Conference (ITSC), 2021 [94].

8. Chapter 12 used materials copyrighted by IEEE. Specifically, the definitive version appeared in the proceedings of the IEEE Transactions on Intelligent Transportation Systems, March 2022 [68].
9. Chapter 13 used materials published by MDPI. Specifically, the definitive version appeared in the proceedings of the Applied Sciences, Applied Sciences, 2021 [69].
10. Chapter 15 used materials copyrighted by IEEE. Specifically, the definitive version appeared in the proceedings of the IEEE 23rd International Conference on Intelligent Transportation Systems (ITSC), 2020 [93].

References

1. Rakesh Agrawal, Tomasz Imieliński, and Arun Swami. Mining association rules between sets of items in large databases. In *Proceedings of the 1993 ACM SIGMOD International Conference on Management of Data*, SIGMOD '93, page 207–216, New York, NY, USA, 1993. Association for Computing Machinery.
2. Faisal Ahmed and Yaser E. Hawas. A threshold-based real-time incident detection system for urban traffic networks. *Procedia - Social and Behavioral Sciences*, 48:1713 – 1722, 2012. Transport Research Arena 2012.
3. M. Arif, G. Wang, and S. Chen. Deep Learning with Non-parametric Regression Model for Traffic Flow Prediction. In *2018 IEEE 16th Intl Conf on Dependable, Autonomic and Secure Computing, 16th Intl Conf on Pervasive Intelligence and Computing, 4th Intl Conf on Big Data Intelligence and Computing and Cyber Science and Technology Congress(DASC/PiCom/DataCom/CyberSciTech)*, pages 681–688, 2018.
4. K. Balke, C. L. Dudek, and C. E. Mountain. Using probe-measured travel time to detect major freeway incidents in houston, texas. *Transportation Research Record, No*, 1554:213–220, 1996.
5. J Barcelo, JF Ferrer, D García, M Florian, and E Le Saux. The parallelization of aimsun2 microscopic simulator for its applications. In *Proceedings of the 3rd. World Congress on Intelligent Transport Systems, Orlando*, 1996.
6. Jaime Barceló and Jordi Casas. Dynamic network simulation with aimsun. In *Simulation approaches in transportation analysis*, pages 57–98. Springer, 2005.
7. Mikhail Belkin and Partha Niyogi. Laplacian eigenmaps for dimensionality reduction and data representation. *Neural Computation*, 15(6):1373–1396, 2003.
8. Christopher M Bishop. *Pattern recognition and machine learning*. springer, 2006.
9. Christian Bliek, Pierre Bonami, and Andrea Lodi. Solving mixed-integer quadratic programming problems with ibm-cplex: a progress report. In *Proceedings of the twenty-sixth RAMP symposium*, pages 16–17, 2014.
10. J. Brügmann, M. Schreckenberg, and W. Luther. Real-time traffic information system using microscopic traffic simulation. In *2013 8th EUROSIM Congress on Modelling and Simulation*, pages 448–453, 2013.
11. Gordon DB Cameron and Gordon ID Duncan. Paramics—parallel microscopic simulation of road traffic. *The Journal of Supercomputing*, 10(1):25–53, 1996.
12. Mustafa Ridvan Cantas and Levent Guvenc. Customized co-simulation environment for autonomous driving algorithm development and evaluation. Technical report, SAE Technical Paper, 2021.
13. Jordi Casas, Jaime L Ferrer, David Garcia, Josep Perarnau, and Alex Torday. Traffic simulation with aimsun. In *Fundamentals of traffic simulation*, pages 173–232. Springer, 2010.
14. Ines Chami, Sami Abu-El-Haija, Bryan Perozzi, Christopher Ré, and Kevin Murphy. Machine learning on graphs: A model and comprehensive taxonomy. *Journal of Machine Learning Research*, 23(89):1–64, 2022.
15. Qianwen Chao, Huikun Bi, Weizi Li, Tianlu Mao, Zhaoqi Wang, and Ming Lin. A survey on visual traffic simulation: Models, evaluations, and applications in autonomous driving. *Computer Graphics Forum*, 39, 07 2019.

16. Sneha Chaudhari, Varun Mithal, Gungor Polatkan, and Rohan Ramanath. An attentive survey of attention models. *ACM Transactions on Intelligent Systems and Technology (TIST)*, 12(5):1–32, 2021.
17. Anthony Chen, Piya Chootinan, Seungkyu Ryu, Ming Lee, and Will Recker. An intersection turning movement estimation procedure based on path flow estimator. *Journal of Advanced Transportation*, 46:161 – 176, 04 2012.
18. Chao-Hua Chen and Gang-Len Chang. A dynamic real-time incident detection system for urban arterials—system architecture and preliminary results. *In Pacific Rim TransTech Conference:*, Volume I: Advanced Technologies, . ASCE:98–104, 1993.
19. Gang Chen. A gentle tutorial of recurrent neural network with error backpropagation. *arXiv preprint arXiv:1610.02583*, 2016.
20. Hao Chen, Ke Yang, Stefano Giovanni Rizzo, Giovanna Vantini, Phillip Taylor, Xiaosong Ma, and Sanjay Chawla. Qarsumo: A parallel, congestion-optimized traffic simulator. *CoRR*, abs/2010.03289, 2020.
21. R. L. Cheu and S. G. Ritchie. Automated detection of lane-blocking freeway incidents using artificial neural networks. *Transp. Res*, 3(6 Part C):371–388, 1995.
22. Kyunghyun Cho, Bart Van Merriënboer, Caglar Gulcehre, Dzmitry Bahdanau, Fethi Bougares, Holger Schwenk, and Yoshua Bengio. Learning phrase representations using rnn encoder-decoder for statistical machine translation. *arXiv preprint arXiv:1406.1078*, 2014.
23. Eleni Christofa, Konstantinos Ampountolas, and Alexander Skabardonis. Arterial traffic signal optimization: A person-based approach. *Transportation Research Part C: Emerging Technologies*, 66:27–47, 2016.
24. T. Chu, J. Wang, L. Codecà, and Z. Li. Multi-agent deep reinforcement learning for large-scale traffic signal control. *IEEE Transactions on Intelligent Transportation Systems*, 21(3):1086–1095, 2020.
25. Li Chun-Gui, Wang Meng, Sun Zi-Gaung, Lin Feiying, and Zhang Zeng-Fang. Urban traffic signal learning control using fuzzy actor-critic methods. *2009 Fifth International Conference on Natural Computation*, 1:368–372, 2009.
26. Andrew Collette. *Python and HDF5*. O'Reilly, 2013.
27. Christopher Day, T.M. Brennan, Hiromel Premachandra, Alexander Hainen, Stephen Remias, James Sturdevant, Greg Richards, Jason Wasson, and Darcy Bullock. Quantifying benefits of traffic signal retiming. *Transportation Research Record*, page pp, 01 2010.
28. Christopher Day, Darcy Bullock, H Li, Stephen Remias, Alexander Hainen, Richard Freije, A.L. Stevens, James Sturdevant, and T.M. Brennan. Performance measures for traffic signal systems: An outcome-oriented approach. *Joint Transportation Research Program*, page pp, 01 2014.
29. Christopher M. Day, Darcy M. Bullock, Howell Li, Steven Lavrenz, W Benjamin Smith, and James R Sturdevant. Integrating traffic signal performance measures into agency business processes. Technical report, Purdue University, West Lafayette, Indiana, 2016.
30. Christopher M Day, Edward J Smaglik, Darcy M Bullock, and James R Sturdevant. Real-time arterial traffic signal performance measures. *FHWA/IN/JTRP-2008/09*, 2008.
31. Michaël Defferrard, Xavier Bresson, and Pierre Vandergheynst. Convolutional neural networks on graphs with fast localized spectral filtering. In D. Lee, M. Sugiyama, U. Luxburg, I. Guyon, and R. Garnett, editors, *Advances in Neural Information Processing Systems*, volume 29. Curran Associates, Inc., 2016.

32. DLR. Theory/Traffic Simulations. `https://sumo.dlr.de/docs/Theory/Traffic\_Simulations.html`, 2020.
33. Alexey Dosovitskiy, Germán Ros, Felipe Codevilla, Antonio M. López, and Vladlen Koltun. Carla: An open urban driving simulator. *ArXiv*, abs/1711.03938, 2017.
34. Simon FG Ehlers. Traffic Queue Length and Pressure Estimation for Road Networks with Geometric Deep Learning Algorithms. *arXiv preprint arXiv:1905.03889*, 2019.
35. Samah El-Tantawy, Baher Abdulhai, and Hossam Abdelgawad. Multiagent reinforcement learning for integrated network of adaptive traffic signal controllers (marlin-atsc): methodology and large-scale application on downtown toronto. *IEEE Transactions on Intelligent Transportation Systems*, 14(3):1140–1150, 2013.
36. Lily Elefteriadou. *An introduction to traffic flow theory*, volume 84. Springer, 2014.
37. M. Fang, L. Tang, X. Yang, Y. Chen, C. Li, and Q. Li. Ftpg: A fine-grained traffic prediction method with graph attention network using big trace data. *IEEE Transactions on Intelligent Transportation Systems*, pages 1–13, 2021.
38. Sebastian Feld, Christoph Roch, Thomas Gabor, Christian Seidel, Florian Neukart, Isabella Galter, Wolfgang Mauerer, and Claudia Linnhoff-Popien. A hybrid solution method for the capacitated vehicle routing problem using a quantum annealer. *arXiv preprint arXiv:1811.07403*, 2018.
39. Félix-Antoine Fortin, François-Michel De Rainville, Marc-André Gardner Gardner, Marc Parizeau, and Christian Gagné. Deap: Evolutionary algorithms made easy. *The Journal of Machine Learning Research*, 13(1):2171–2175, 2012.
40. Vincent François-Lavet, Peter Henderson, Riashat Islam, Marc G Bellemare, and Joelle Pineau. An introduction to deep reinforcement learning. *arXiv preprint arXiv:1811.12560*, 2018.
41. Richard Freije, Alexander Hainen, Amanda Stevens, Howell Li, W. Smith, Hayley Summers, Christopher Day, James Sturdevant, and Darcy Bullock. Graphical performance measures for practitioners to triage split failure trouble calls. *Transportation Research Record: Journal of the Transportation Research Board*, 2439:27–40, 12 2014.
42. Qiao Ge, Biagio Ciuffo, and Monica Menendez. An exploratory study of two efficient approaches for the sensitivity analysis of computationally expensive traffic simulation models. *IEEE Transactions on Intelligent Transportation Systems*, 15(3):1288–1297, 2014.
43. Qiao Ge and Monica Menendez. An improved approach for the sensitivity analysis of computationally expensive microscopic traffic models: a case study of the zurich network in vissim. In *Transportation Research Board 92nd Annual Meeting*, 2013.
44. Wade Genders and Saiedeh Razavi. Using a deep reinforcement learning agent for traffic signal control. *arXiv preprint arXiv:1611.01142*, 2016.
45. Mohammad Shareef Ghanim and Khaled Shaaban. Estimating turning movements at signalized intersections using artificial neural networks. *IEEE Transactions on Intelligent Transportation Systems*, 20(5):1828–1836, 2018.
46. Fred Glover, Gary Kochenberger, and Yu Du. A tutorial on formulating and using qubo models. *arXiv preprint arXiv:1811.11538*, 2018.
47. Gabriel Gomes, Juliette Ugirumurera, and Xiaoye S. Li. Distributed macroscopic traffic simulation with open traffic models. In *2020 IEEE 23rd International Conference on Intelligent Transportation Systems (ITSC)*, pages 1–6, 2020.
48. Sorin Grigorescu, Bogdan Trasnea, Tiberiu Cocias, and Gigel Macesanu. A survey of deep learning techniques for autonomous driving. *Journal of Field Robotics*, 37(3):362–386, 2020.

49. Yiming Gu, Zhen Sean Qian, and Feng Chen. From twitter to detector: Real-time traffic incident detection using social media data. *Transportation research part C: emerging technologies*, 67:321–342, 2016.
50. Moritz Gutlein, Reinhard German, and Anatoli Djanatliev. Towards a hybrid co-simulation framework: Hla-based coupling of matsim and sumo. In *2018 IEEE/ACM 22nd International Symposium on Distributed Simulation and Real Time Applications (DS-RT)*, pages 1–9, 2018.
51. Abolhassan Halati, Henry Lieu, and Susan Walker. Corsim-corridor traffic simulation model. In *Traffic Congestion and Traffic Safety in the 21st Century: Challenges, Innovations, and OpportunitiesUrban Transportation Division, ASCE; Highway Division, ASCE; Federal Highway Administration, USDOT; and National Highway Traffic Safety Administration, USDOT.*, 1997.
52. William L. Hamilton, Rex Ying, and Jure Leskovec. Inductive representation learning on large graphs. In *Proceedings of the 31st International Conference on Neural Information Processing Systems*, NIPS'17, page 1025–1035, Red Hook, NY, USA, 2017. Curran Associates Inc.
53. Lee D. Han and Adolf Darlington May. Automatic detection of traffic operational problems on urban arterials. *No.*, pages 9–15, 1989.
54. Charles R. Harris, K. Jarrod Millman, St'efan J. van der Walt, Ralf Gommers, Pauli Virtanen, David Cournapeau, Eric Wieser, Julian Taylor, Sebastian Berg, Nathaniel J. Smith, Robert Kern, Matti Picus, Stephan Hoyer, Marten H. van Kerkwijk, Matthew Brett, Allan Haldane, Jaime Fernández del Río, Mark Wiebe, Pearu Peterson, Pierre Gérard-Marchant, Kevin Sheppard, Tyler Reddy, Warren Weckesser, Hameer Abbasi, Christoph Gohlke, and Travis E. Oliphant. Array programming with NumPy. *Nature*, 585(7825):357–362, September 2020.
55. William E Hart, Carl D Laird, Jean-Paul Watson, David L Woodruff, Gabriel A Hackebeil, Bethany L Nicholson, and John D Siirola. *Pyomo-optimization modeling in python*, volume 67. Springer, 2017.
56. Tingting Huang, Subhadipto Poddar, Cristopher Aguilar, Anuj Sharma, Edward Smaglik, Sirisha Kothuri, and Peter Koonce. Building intelligence in automated traffic signal performance measures with advanced data analytics. *Transportation Research Record*, 2672(18):154–166, 2018.
57. Hasham Hussain, Muhammad bin Javaid, Faisal Shah Khan, Aeysha Khalique, and Archismita Dalal. Optimal Control of Traffic Signals using Quantum Annealing. *arXiv preprint arXiv:1912.07134*, 2019.
58. Daisuke Inoue, Akihisa Okada, Tadayoshi Matsumori, Kazuyuki Aihara, and Hiroaki Yoshida. Traffic Signal Optimization on a Square Lattice using the D-Wave Quantum Annealer. *arXiv preprint arXiv:2003.07527*, 2020.
59. INRIX. Los angeles tops inrix global congestion ranking. `https://inrix.com/press-releases/scorecard-2017/`, 2018. (Accessed on 7/20/2021).
60. Young-Seon Jeong, Manoel Castro-Neto, Myong K. Jeong, and Lee D. Han. A wavelet-based freeway incident detection algorithm with adapting threshold parameters. *Transportation Research Part C: Emerging Technologies*, 19(1):1–19, 2011.
61. Yangsheng Jiang, Zhihong Yao, Xiaoling Luo, Weitiao Wu, Xiao Ding, and Afaq Khattak. Heterogeneous platoon flow dispersion model based on truncated mixed simplified phase-type distribution of travel speed. *Journal of Advanced Transportation*, 50(8):2160–2173, 2016.

62. Junchen Jin and Xiaoliang Ma. Adaptive group-based signal control using reinforcement learning with eligibility traces. In *2015 IEEE 18th International Conference on Intelligent Transportation Systems*, pages 2412–2417. IEEE, 2015.
63. Xin Jin, Ruey Long Cheu, and Dipti Srinivasan. Development and adaptation of constructive probabilistic neural network in freeway incident detection. *Transportation Research Part C: Emerging Technologies*, 10(2):121–147, 2002.
64. Yashaswi Karnati., Dhruv Mahajan., Anand Rangarajan., and Sanjay Ranka. Data mining algorithms for traffic interruption detection. In *Proceedings of the 6th International Conference on Vehicle Technology and Intelligent Transport Systems - Volume 1: VEHITS,*, pages 106–114. INSTICC, SciTePress, 2020.
65. Yashaswi Karnati, Dhruv Mahajan, Anand Rangarajan, and Sanjay Ranka. Machine learning algorithms for traffic interruption detection. In *2020 Fifth International Conference on Fog and Mobile Edge Computing (FMEC)*, pages 231–236, 2020.
66. Yashaswi Karnati, Rahul Sengupta, Anand Rangarajan, and Sanjay Ranka. Subcycle-based neural network algorithms for turning movement count prediction. In *International Conference on Vehicle Technology and Intelligent Transport Systems*, 2021.
67. Yashaswi Karnati, Rahul Sengupta, Anand Rangarajan, and Sanjay Ranka. Subcycle waveform modeling of traffic intersections using recurrent attention networks. *IEEE Transactions on Intelligent Transportation Systems*, 23(3):2538–2548, 2021.
68. Yashaswi Karnati, Rahul Sengupta, Anand Rangarajan, and Sanjay Ranka. Subcycle waveform modeling of traffic intersections using recurrent attention networks. *IEEE Transactions on Intelligent Transportation Systems*, 23(3):2538–2548, 2022.
69. Yashaswi Karnati, Rahul Sengupta, and Sanjay Ranka. Intertwin: Deep learning approaches for computing measures of effectiveness for traffic intersections. *Applied Sciences*, 11(24):11637, Dec 2021.
70. M. F. Kasim, D. Watson-Parris, L. Deaconu, S. Oliver, P. Hatfield, D. H. Froula, G. Gregori, M. Jarvis, S. Khatiwala, J. Korenaga, J. Topp-Mugglestone, E. Viezzer, and S. M. Vinko. Building high accuracy emulators for scientific simulations with deep neural architecture search, 2020.
71. Khewal Bhupendra Kesur. Advances in genetic algorithm optimization of traffic signals. *Journal of Transportation Engineering*, 135(4):160–173, 2009.
72. Mohamed A Khamis and Walid Gomaa. Adaptive multi-objective reinforcement learning with hybrid exploration for traffic signal control based on cooperative multi-agent framework. *Engineering Applications of Artificial Intelligence*, 29:134–151, 2014.
73. Thomas N. Kipf and Max Welling. Semi-supervised classification with graph convolutional networks. In *5th International Conference on Learning Representations, ICLR 2017, Toulon, France, April 24-26, 2017, Conference Track Proceedings*. OpenReview.net, 2017.
74. R. Klefstad, Yue Zhang, Mingjie Lai, R. Jayakrishnan, and R. Lavanya. A distributed, scalable, and synchronized framework for large-scale microscopic traffic simulation. In *Proceedings. 2005 IEEE Intelligent Transportation Systems, 2005.*, pages 813–818, 2005.
75. Slawomir Koziel and Xin-She Yang. *Computational optimization, methods and algorithms*, volume 356. Springer, 2011.
76. Charles R. Lattimer. *Automated Traffic Signal Performance Measures (ATSPMs)*. FHWA, Washington, D.C., 2020.

77. Jung-Taek Lee and William C. Taylor. Application of a dynamic model for arterial street incident detection. *ITS Journal - Intelligent Transportation Systems Journal*, 5(1):53–70, 1999.
78. Florin Leon, Andrei-Stefan Lupu, and Costin Badica. Multiagent coalition structure optimization by quantum annealing. In *Computational Collective Intelligence: 9th International Conference, ICCCI 2017, Nicosia, Cyprus, September 27-29, 2017, Proceedings, Part I 9*, pages 331–341. Springer, 2017.
79. R. Levie, F. Monti, X. Bresson, and M. M. Bronstein. Cayleynets: Graph convolutional neural networks with complex rational spectral filters. *IEEE Transactions on Signal Processing*, 67(1):97–109, 2019.
80. Howell Li, Lucy M. Richardson, Christopher Day, James Howard, and Darcy Bullock. Scalable dashboard for identifying split failures and heuristic for reallocating split times. *Transportation Research Record: Journal of the Transportation Research Board*, 2620:83–95, 01 2017.
81. Wenqing Li, Chuhan Yang, and Saif Eddin Jabari. Short-term traffic forecasting using high-resolution traffic data. In *2020 IEEE 23rd International Conference on Intelligent Transportation Systems (ITSC)*. IEEE, sep 2020.
82. Yuxi Li. Deep reinforcement learning: An overview. *arXiv preprint arXiv:1701.07274*, 2017.
83. M. J. Lighthill and G. B. Whitham. On kinematic waves. II. A theory of traffic flow on long crowded roads. *Proc. Roy. Soc. London Ser. A*, 229:317–345, 1955.
84. Wei-Hua Lin and Carlos F. Daganzo. A simple detection scheme for delay-inducing freeway incidents. *Transportation Research Part A: Policy and Practice*, 31(2):141 – 155, 1997.
85. John Little, Mark Kelson, and Nathan Gartner. Maxband: A program for setting signals on arteries and triangular networks. *Transportation Research Record Journal of the Transportation Research Board*, 795:40–46, 12 1981.
86. Henry X Liu and Wenteng Ma. A virtual vehicle probe model for time-dependent travel time estimation on signalized arterials. *Transportation Research Part C: Emerging Technologies*, 17(1):11–26, 2009.
87. Weiwei Liu, Xichen Wang, Wenli Zhang, Lin Yang, and Chao Peng. Coordinative simulation with sumo and ns3 for vehicular ad hoc networks. In *2016 22nd Asia-Pacific Conference on Communications (APCC)*, pages 337–341, 2016.
88. Pablo Alvarez Lopez, Michael Behrisch, Laura Bieker-Walz, Jakob Erdmann, Yun-Pang Flotterod, Robert Hilbrich, Leonhard Lucken, Johannes Rummel, Peter Wagner, and Evamarie Wiessner. Microscopic traffic simulation using sumo. In *The 21st IEEE International Conference on Intelligent Transportation Systems*. IEEE, 2018.
89. Nicholas E Lownes and Randy B Machemehl. Vissim: a multi-parameter sensitivity analysis. In *Proceedings of the 2006 Winter Simulation Conference*, pages 1406–1413. IEEE, 2006.
90. Y. Lv, Y. Duan, W. Kang, Z. Li, and F. Wang. Traffic flow prediction with big data: A deep learning approach. *IEEE Transactions on Intelligent Transportation Systems*, 16(2):865–873, April 2015.
91. Dhruv Mahajan, Tania Banerjee, Yashaswi Karnati, Anand Rangarajan, and Sanjay Ranka. A data driven approach to derive traffic intersection geography using high resolution controller logs. In *VEHITS*, pages 203–210, 01 2020.

92. Dhruv Mahajan, Tania Banerjee, Anand Rangarajan, Nithin Agarwal, Jeremy Dilmore, Emmanuel Posadas, and Sanjay Ranka. Analyzing traffic signal performance measures to automatically classify signalized intersections. In *VEHITS*, pages 138–147, 2019.
93. Dhruv Mahajan, Yashaswi Karnati, Tania Banerjee, Varun Reddy Regalla, Rohit Reddy, Anand Rangarajan, and Sanjay Ranka. A scalable data analytics and visualization system for city-wide traffic signal data-sets. In *2020 IEEE 23rd International Conference on Intelligent Transportation Systems (ITSC)*, pages 1–6, 2020.
94. Dhruv Mahajan, Yashaswi Karnati, Jeremy Dilmore, Anand Rangarajan, and Sanjay Ranka. An automated framework for deriving intersection coordination plans. In *ITSC*, pages 1322–1327, 2021.
95. Dhruv Mahajan, Yashaswi Karnati, Anand Rangarajan, and Sanjay Ranka. Unsupervised summarization and change detection in high-resolution signalized intersection datasets. In *2020 IEEE 23rd International Conference on Intelligent Transportation Systems (ITSC)*, pages 1–6, 2020.
96. A. Makhorin. Glpk (gnu linear programming kit). Available at `http://www.gnu.org/software/glpk/glpk.html`.
97. Meher P Malakapalli and Carroll J Messer. Enhancements to the passer ii-90 delay estimation procedures. *Transportation Research Record*, pages 94–103, 1993.
98. Catherine C McGeoch. Theory versus practice in annealing-based quantum computing. *Theoretical Computer Science*, 816:169–183, 2020.
99. Juan C Medina and Rahim F Benekohal. Traffic signal control using reinforcement learning and the max-plus algorithm as a coordinating strategy. In *2012 15th International IEEE Conference on Intelligent Transportation Systems*, pages 596–601. IEEE, 2012.
100. K. C. Mouskos, E. Niver, S. Lee, T. Batz, and P. Dwyer. Transportation operation coordinating committee system for managing incidents and traffic: evaluation of the incident detection system. *Transportation Research Record, No*, 1679:50–57, 1999.
101. Kai Nagel and Marcus Rickert. Parallel implementation of the transims micro-simulation. *Parallel Computing*, 27:1611–1639, 11 2001.
102. Florian Neukart, Gabriele Compostella, Christian Seidel, David Von Dollen, Sheir Yarkoni, and Bob Parney. Traffic flow optimization using a quantum annealer. *Frontiers in ICT*, 4:29, 2017.
103. Nga TT Nguyen, Garrett T Kenyon, and Boram Yoon. A regression algorithm for accelerated lattice qcd that exploits sparse inference on the d-wave quantum annealer. *Scientific Reports*, 10(1):1–7, 2020.
104. OpenStreetMap contributors. Planet dump retrieved from https://planet.osm.org . `https://www.openstreetmap.org`, 2017.
105. Kaan Ozbay, Hong Yang, ENDER FARUK Morgul, Sandeep Mudigonda, and Bekir Bartin. Big data and the calibration and validation of traffic simulation models. In *Traffic and Transportation Simulation-Looking Back and Looking Ahead: Celebrating 50 Years of Traffic Flow Theory, a Workshop*, pages 92–122. Transportation Research Board Washington, DC, 2014.
106. M. C. Bell P. T. Martin. Network programming to derive turning movements from link flows. In *Transp. Res. Rec., no. 1365*, pages 147–154. Transportation Research Board, 1992.
107. GM Pacey. The progress of a bunch of vehicles released from a traffic signal. research note rn/2665/gmp. road research laboratory, 1956.

108. Hyoshin Park and Ali Haghani. Real-time prediction of secondary incident occurrences using vehicle probe data. *Transportation Research Part C: Emerging Technologies*, 70:69–85, 2016.
109. Emily Parkany and Chi Xie. A complete review of incident detection algorithms & their deployment: what works and what doesn't. *Transportation Research Records*, 2005.
110. Adam Paszke, Sam Gross, Francisco Massa, Adam Lerer, James Bradbury, Gregory Chanan, Trevor Killeen, Zeming Lin, Natalia Gimelshein, Luca Antiga, Alban Desmaison, Andreas Kopf, Edward Yang, Zachary DeVito, Martin Raison, Alykhan Tejani, Sasank Chilamkurthy, Benoit Steiner, Lu Fang, Junjie Bai, and Soumith Chintala. PyTorch: An Imperative Style, High-Performance Deep Learning Library, 2019.
111. Lang Yang Pengpeng Jiao, Huapu Lu. Real-time estimation of turning movement proportions based on genetic algorithm. In *Proceedings. 2005 IEEE Intelligent Transportation Systems, 2005.*, pages 96–101, Sep. 2005.
112. Mahmoud Pourmehrab, Lily Elefteriadou, and Sanjay Ranka. Real-time intersection optimization for signal phasing, timing, and automated vehicles' trajectories. *arXiv preprint arXiv:2007.03763*, 2020.
113. Kotagiri Ramamohanarao, Hairuo Xie, Lars Kulik, Shanika Karunasekera, Egemen Tanin, Rui Zhang, and Eman Bin Khunayn. Smarts: Scalable microscopic adaptive road traffic simulator. *ACM Trans. Intell. Syst. Technol.*, 8(2), dec 2016.
114. Dennis I Robertson. Transyt: a traffic network study tool, 1969.
115. Matthew Rocklin. Dask: Parallel computation with blocked algorithms and task scheduling. In Kathryn Huff and James Bergstra, editors, *Proceedings of the 14th Python in Science Conference*, pages 130–136, 2015.
116. Alvaro Sanchez-Gonzalez, Jonathan Godwin, Tobias Pfaff, Rex Ying, Jure Leskovec, and Peter Battaglia. Learning to simulate complex physics with graph networks. In *International Conference on Machine Learning*, pages 8459–8468. PMLR, 2020.
117. A. Saroj, S. Roy, A. Guin, M. Hunter, and R. Fujimoto. Smart city real-time data-driven transportation simulation. In *2018 Winter Simulation Conference (WSC)*, pages 857–868, 2018.
118. Roeland Scheepens, Christophe Hurter, Huub Van De Wetering, and Jarke J Van Wijk. Visualization, selection, and analysis of traffic flows. *IEEE transactions on visualization and computer graphics*, 22(1):379–388, 2015.
119. B Seibold. A mathematical introduction to traffic flow theory. *IPAM Tutorials*, 2015.
120. M. W. Sermons and F. S. Koppelman. Use of vehicle positioning data for arterial incident detection. *Transportation Research Part C*, 4(2):87–96, 1996.
121. Colin Sheppard, Rashid Waraich, Andrew Campbell, Alexei Pozdnukov, and Anand R Gopal. Modeling plug-in electric vehicle charging demand with beam: The framework for behavior energy autonomy mobility. Technical report, Lawrence Berkeley National Lab.(LBNL), Berkeley, CA (United States), 2017.
122. Christoph Sommer, David Eckhoff, Alexander Byers Brummer, Dominik S. Buse, Florian Hagenauer, Stefan Joerer, and Michele Segata. Veins: The open source vehicular network simulation framework. *Recent Advances in Network Simulation*, 2019.
123. Chronis Stamatiadis and Nathan H Gartner. Multiband-96: a program for variable-bandwidth progression optimization of multiarterial traffic networks. *Transportation Research Record*, 1554(1):9–17, 1996.

124. Tobias Stollenwerk, Bryan O'Gorman, Davide Venturelli, Salvatore Mandrà, Olga Rodionova, Hokkwan Ng, Banavar Sridhar, Eleanor Gilbert Rieffel, and Rupak Biswas. Quantum annealing applied to de-conflicting optimal trajectories for air traffic management. *IEEE transactions on intelligent transportation systems*, 21(1):285–297, 2019.
125. Jian Sun and Lun Zhang. Vehicle actuation based short-term traffic flow prediction model for signalized intersections. *Journal of Central South University*, 19(1):287–298, 2012.
126. Richard S Sutton and Andrew G Barto. *Reinforcement learning: An introduction*. MIT press, 2018.
127. Pete Sykes. Traffic simulation with paramics. In *Fundamentals of traffic simulation*, pages 131–171. Springer, 2010.
128. Hualiang Teng and Yi Qi. Application of wavelet technique to freeway incident detection. *Transportation Research Part C: Emerging Technologies*, 11(3-4):289–308, 2003.
129. S. Thancanamootoo and Matthew G. H. Bell. Automatic detection of traffic incidents on a signal-controlled road network. *Research Report*, 76, 1988.
130. Sunil Thulasidasan and Stephan Eidenbenz. Accelerating traffic microsimulations: A parallel discrete-event queue-based approach for speed and scale. In *Proceedings of the 2009 Winter Simulation Conference (WSC)*, pages 2457–2466, 2009.
131. Ledyard R. Tucker. Some mathematical notes on three-mode factor analysis. *Psychometrika*, 31(3):279–311, 1966.
132. UDOT. UDOT automated traffic signal performance measures - automated traffic signal performance metrics. `https://udottraffic.utah.gov/atspm/`, 2017. (Accessed on 2/26/2020).
133. Federal Highway Administration US Department of Transportation. Traffic signal timing manual. `https://ops.fhwa.dot.gov/publications/fhwahop08024/chapter4.htm`, 06 2008. (Accessed on 03/18/2021).
134. Federal Highway Administration US Department of Transportation. Traffic signal timing manual. `https://ops.fhwa.dot.gov/publications/fhwahop08024/index.htm`, 06 2008. (Accessed on 2/10/2020).
135. Federal Highway Administration US Department of Transportation. Measures of effectiveness and validation guidance for adaptive signal control technologies. `https://ops.fhwa.dot.gov/publications/fhwahop13031/index.htm`, 07 2013. (Accessed on 2/10/2020).
136. US DOT FHWA. Types of Traffic Analysis Tools. `https://ops.fhwa.dot.gov/trafficanalysistools/type\_tools.htm`, 2020.
137. Elise Van der Pol and Frans A Oliehoek. Coordinated deep reinforcement learners for traffic light control. *Proceedings of learning, inference and control of multi-agent systems (at NIPS 2016)*, 8:21–38, 2016.
138. Ashish Vaswani, Noam Shazeer, Niki Parmar, Jakob Uszkoreit, Llion Jones, Aidan N Gomez, Lukasz Kaiser, and Illia Polosukhin. Attention is all you need. *Advances in neural information processing systems*, 30, 2017.
139. Shuling Wang, Wei Huang, and Hong K Lo. Traffic parameters estimation for signalized intersections based on combined shockwave analysis and Bayesian Network. *Transportation Research, Part C: Emerging Technologies*, 104:22–37, 2019.
140. Yabo Wang, Shuning Xu, and Di Feng. A New Method for Short-term Traffic Flow Prediction Based on Multi-segments Features. In *Proceedings of the 2020 12th International Conference on Machine Learning and Computing*, pages 34–38, 2020.

141. Yanpeng Wang, Leina Zhao, Shuqing Li, Xinyu Wen, and Yang Xiong. Short Term Traffic Flow Prediction of Urban Road Using Time Varying Filtering Based Empirical Mode Decomposition. *Applied Sciences*, 10(6):2038, 2020.
142. Yuan Wang, Dongxiang Zhang, Ying Liu, Bo Dai, and Loo Hay Lee. Enhancing transportation systems via deep learning: A survey. *Transportation research part C: emerging technologies*, 99:144–163, 2019.
143. Yuan Wang, Dongxiang Zhang, Ying Liu, Bo Dai, and Loo Hay Lee. Enhancing transportation systems via deep learning: A survey. *Transportation research part C: emerging technologies*, 99:144–163, 2019.
144. Yuan Wang, Dongxiang Zhang, Ying Liu, Bo Dai, and Loo Hay Lee. Enhancing transportation systems via deep learning: A survey. *Transportation research part C: emerging technologies*, 99:144–163, 2019.
145. Fo Vo Webster. Traffic signal settings, 1958.
146. Hua Wei, Guanjie Zheng, Vikash V. Gayah, and Zhenhui Li. A survey on traffic signal control methods. *CoRR*, abs/1904.08117, 2019.
147. T. D. Wemegah and S. Zhu. Big data challenges in transportation: A case study of traffic volume count from massive radio frequency identification(rfid) data. In *2017 International Conference on the Frontiers and Advances in Data Science (FADS)*, pages 58–63, Oct 2017.
148. B. M. Williams and A. Guin. Traffic management center use of incident detection algorithms: Findings of a nationwide survey. *Trans. Intell. Transport. Sys.*, 8(2):351–358, June 2007.
149. Ronald J. Williams and David Zipser. A learning algorithm for continually running fully recurrent neural networks. *Neural Computation*, 1(2):270–280, 1989.
150. SC Wong, WT Wong, CM Leung, and CO Tong. Group-based optimization of a time-dependent transyt traffic model for area traffic control. *Transportation Research Part B: Methodological*, 36(4):291–312, 2002.
151. Matthew A Wright, Simon FG Ehlers, and Roberto Horowitz. Neural-Attention-Based Deep Learning Architectures for Modeling Traffic Dynamics on Lane Graphs. In *2019 IEEE Intelligent Transportation Systems Conference (ITSC)*, pages 3898–3905, 2019.
152. J. Wu and C. Thnay. An o-d based method for estimating link and turning volume based on counts. In *Transp. Res. Rec., no. 1365*, pages 865–873. ITE Dist 6, 2001.
153. Weitiao Wu, Wenzhou Jin, and Luou Shen. Mixed platoon flow dispersion model based on speed-truncated gaussian mixture distribution. *Journal of Applied Mathematics*, 2013:480965, Jun 2013.
154. Weitiao Wu, Luou Shen, Wenzhou Jin, and Ronghui Liu. Density-based mixed platoon dispersion modelling with truncated mixed gaussian distribution of speed. *Transportmetrica B: Transport Dynamics*, 3(2):114–130, 2015.
155. Dongwei Xu, Chenchen Wei, Peng Peng, Qi Xuan, and Haifeng Guo. GE-GAN: A novel deep learning framework for road traffic state estimation. *Transportation Research, Part C: Emerging Technologies*, 117:102635, 2020.
156. Kun Xu, Ping Yi, Chun Shao, and Jialei Mao. Development and testing of an automatic turning movement identification system at signalized intersections. *Journal of Transportation Technologies*, 03:241–246, 01 2013.
157. Runsheng Xu, Yi Guo, Xu Han, Xin Xia, Hao Xiang, and Jiaqi Ma. Opencda: An open cooperative driving automation framework integrated with co-simulation. In *2021 IEEE International Intelligent Transportation Systems Conference (ITSC)*, pages 1155–1162, 2021.

158. Hong Yang, Zhenyu Wang, Kun Xie, and Dong Dai. Use of ubiquitous probe vehicle data for identifying secondary crashes. *Transportation research part C: emerging technologies*, 82:138–160, 2017.
159. Kok-Lim Alvin Yau, Junaid Qadir, Hooi Ling Khoo, Mee Hong Ling, and Peter Komisarczuk. A survey on reinforcement learning models and algorithms for traffic signal control. *ACM Computing Surveys (CSUR)*, 50(3):1–38, 2017.
160. Bing Yu, Haoteng Yin, and Zhanxing Zhu. Spatio-temporal graph convolutional networks: A deep learning framework for traffic forecasting. In *Proceedings of the Twenty-Seventh International Joint Conference on Artificial Intelligence, IJCAI-18*, pages 3634–3640. International Joint Conferences on Artificial Intelligence Organization, 7 2018.
161. Shahriar Afandizadeh Zargari, Salar Zabihi Siabil, Amir Hossein Alavi, and Amir Hossein Gandomi. A computational intelligence-based approach for short-term traffic flow prediction. *Expert Systems*, 29(2):124–142, 2012.
162. Huichu Zhang, Siyuan Feng, Chang Liu, Yaoyao Ding, Yichen Zhu, Zihan Zhou, Weinan Zhang, Yong Yu, Haiming Jin, and Zhenhui Li. Cityflow: A multi-agent reinforcement learning environment for large scale city traffic scenario. *CoRR*, abs/1905.05217, 2019.
163. D. Zhao, Y. Dai, and Z. Zhang. Computational Intelligence in Urban Traffic Signal Control: A Survey. *IEEE Transactions on Systems, Man, and Cybernetics, Part C (Applications and Reviews)*, 42(4):485–494, 2012.
164. L. Zhao, Y. Song, C. Zhang, Y. Liu, P. Wang, T. Lin, M. Deng, and H. Li. T-gcn: A temporal graph convolutional network for traffic prediction. *IEEE Transactions on Intelligent Transportation Systems*, 21(9):3848–3858, 2020.
165. Zhi-Yun Zou, Shao-Kuan Chen, Jin-yi GUO, Li-qiong BAI, and Chao-fan CHANG. Timing optimization and simulation on signalized intersection by synchro [j]. *Journal of Northern Jiaotong University*, 6, 2004.

Index

Printed in the United States
by Baker & Taylor Publisher Services